From Melting Ice to Rising Seas

From Melting Ice to Rising Seas

Global Warming's Effects

RAFEAL MECHLORE

Readers Publications

CONTENTS

CONTENTS

INDEX

Introduction

Chapter 6 : Rising Sea Levels
6.1 Discussion on sea level rise and its measurements
6.2 The link between melting ice and rising seas
6.3 Coastal regions and cities most vulnerable to sea level rise
Chapter 7 : Ocean Acidification
7.1 Explanation of the ocean's absorption of excess CO_2
7.2 Impacts on marine life, including coral reefs and shellfish
7.3 Global consequences for fisheries and food security

Chapter 8 : Extreme Weather Events
8.1 Examination of the relationship between global warming and extreme weather
8.2 Climate adaptation and disaster management

Chapter 9 : Socioeconomic Implications
9.1 The human toll of global warming, including displacement and health effects
9.2 Economic consequences of climate change and mitigation efforts
9.3 International agreements and cooperation on climate action

Chapter 10 : Mitigation and Adaptation
10.1 Strategies for reducing greenhouse gas emissions
10.2 Approaches to mitigating global warming's effects on ice and seas
10.3 Efforts to adapt to a changing climate, both locally and globally

Chapter 11 : Hope and Solutions
11.1 Success stories of communities and nations taking action
11.2 Innovations in renewable energy and climate-friendly technology
11.3 The role of individual and collective action in combating global warming

Chapter 12 : Looking to the Future
12.1 Anticipating the long-term consequences of melting ice and rising seas
12.2 The urgency of addressing global warming for future generations
12.3 Call to action and a hopeful vision for a sustainable future

INTRODUCTION

The planet we call home is going through a dramatic change at the moment, one that is being driven by the unrelenting force of global warming. The results of our activities are becoming more and more obvious to us as we watch the ice melt and see the sea levels rise. The world is at a crossroads, confronting a crisis that threatens not just the delicate ecosystems on our planet but also the very foundations of human civilization. This catastrophe is a result of global warming, which is a result of human activity.

The documentary "From Melting Ice to Rising Seas: Global Warming's Effects" begins on a journey to unravel the intricate fabric of our planet's changing climate. This book is a testament to the importance of comprehending the social, economic, and environmental ramifications of global warming as quickly as possible. It acts as an exhaustive guide to the undiscovered frontiers of our warming world, providing insights into the deep transformations that are taking place in the Earth's cryosphere, the relentless rise of sea levels, and the wider ramifications for all species on this blue globe that we call home.

The repercussions of global warming have transcended scientific jargon and climate conferences to become a defining concern of our day. This is because we live in an era of unprecedented scientific understanding and technological innovation. Every glacier that melts, every heatwave that sets a new record, and every coastal city that is threatened by increasing seas serve as a stark reminder of the urgent need to face this crisis head-on.

The first step in the process is gaining a grasp of global warming in general, as well as the underlying factors that contribute to it and the unstoppable increase of greenhouse gases in our atmosphere. We go into the scientific literature in an effort to untangle the complex web of the greenhouse effect, which is the driving force behind the changing climate. We investigate the historical context, tracking the findings and the road that brought us to the realization that global warming is one of the most critical problems humanity currently faces, and we discuss how this realization came about.

The cryosphere, a region of the Earth that includes polar ice caps, glaciers, and permafrost, is the primary focus of the book and serves as its central theme. These frozen zones, which are frequently neglected in more general debates about climate, are essential to the process of controlling the temperature of our globe and play an important part in preserving the equilibrium of the Earth's climate.

As we travel throughout these frozen landscapes, we are gradually gaining a better understanding of the cryosphere and the role it plays within the larger climate system.

In the following pages, we will begin an investigation of the phenomena of the melting of the polar ice caps, which has drawn attention all across the world. Our journey takes us to the Arctic and Antarctic regions, two places on the planet where the effects of a warming climate are most readily apparent. As we learn more about these areas, we are getting a better understanding of the causes and impacts of the melting of polar ice, as well as the implications that this phenomenon has for ecosystems, native communities, and the larger global climate system.

Our travels lead us to the receding glaciers of the world, where we investigate regional case studies in regions such as the Himalayas, Andes, and Alps. The melting of glaciers is not only a striking illustration of the way in which our planet is transforming, but it also serves as a useful signal of the difficulties that lie ahead. We make the connection between the melting of glaciers, the depletion of freshwater resources, and the impact that this has on the local economy and communities in these crucial areas.

The thawing of permafrost, another aspect of climate change that is sometimes overlooked, is the focus of our undivided attention. As we continue to investigate this topic, we are gradually uncovering the mysteries of permafrost, which is a frozen repository of ancient carbon. We take a close look at the process by which greenhouse gases are released as a result of thawing permafrost and the far-reaching repercussions that this process has, in particular for the infrastructure and landscapes of the Arctic and other regions.

Now that we have established the cryosphere as an essential component of our investigation, we will investigate the unstoppable rise in sea levels. We investigate the complexities of rising sea levels and assess the effects that this phenomenon will have on coastal areas and the cities that are most at risk from the advancing waters. We also address ocean acidification, which is a less well-known aspect of global warming, as well as the effects it has on marine life, coral reefs, and fisheries around the world.

The effects of global warming on the environment are only one topic that is covered in this book; it also dives deeply into the social repercussions that will result from these shifts. As we progress through the chapters, we will reveal the human toll, which will include the effects of climate change on things like relocation, health, and the far-reaching economic repercussions of this phenomenon. We take a look at the global initiatives for reducing emissions and increasing resilience to climate change, highlighting the important part that individuals, communities, and governments all play in finding solutions to the climate challenge.

Despite this, there is more to the story than just the obstacles. In the book "From Melting Ice to Rising Seas," we present stories of optimism and potential solutions to the problem of rising sea levels. We shine a focus on nations and communities that have already begun taking steps and initiating projects that point the way toward a more sustainable future. We investigate recent advancements in climate-friendly technologies and renewable energy sources, shedding light on the way forward for a global community working together to combat the effects of global warming.

As we continue to read through these pages, we come to the realization that the urgency of tackling global warming is not only a concern for the environment; rather, it is an existential and moral necessity. The decisions that we make right now will reverberate through the annals of history and help shape the world that we leave for the generations to come. The book "From Melting Ice to Rising Seas: The Effects of Global Warming" is an invitation to engage with the issues of our day, a call to action, and a plea to acknowledge the profound interconnection of our world.

We have written this book in the hopes that it will spark a collective awakening and motivate all of us to become stewards of the Earth. We must all work together to counteract the impacts of global warming and ensure that our children and their descendants will inherit a future that is both sustainable and habitable. The voyage starts right here, at the center of our rapidly transforming world, where the melting ice caps meet the expanding oceans.

1. **Overview of global warming and its causes**

    The phenomenon known as global warming, which is characterized by the progressive increase in the average temperature of the Earth, has emerged as the preeminent environmental as well as socioeconomic problem of our generation. The effects of global warming are being felt all around the planet as the global climate system goes through considerable adjustments. These effects are having an effect on ecosystems, economies, and the way people make their livings. It is absolutely necessary to have a solid understanding of the complexities of global warming and the many different factors that contribute to it in order to be able to devise effective methods to reduce the negative effects of this phenomenon and protect the future of our planet.

    The Science That Supports the Hypothesis of Global Warming:

    The greenhouse effect is at the heart of the problem of global warming. This is a naturally occurring process that maintains a constant temperature on Earth by retaining heat in the atmosphere. As greenhouse gases, substances like carbon dioxide (CO2), methane (CH4), and water vapor allow sunlight to reach the surface of the Earth while preventing heat from escaping into space. While these gases allow sunlight to reach the surface of the Earth, they also prevent heat from escaping into space. This natural mechanism is very necessary in order to keep the temperature of the planet within the parameters of what is considered to be habitable.

    The effects of human activity are:

Even though the greenhouse effect is a naturally occurring phenomenon, the consequences of it have been considerably exacerbated as a result of human activity. The release of an excessive amount of greenhouse gases into the atmosphere has been contributed to by the burning of fossil fuels, the destruction of forests, industrial operations, and agricultural activities. This has had the effect of amplifying the planet's inherent greenhouse effect. The extraordinary increase in the concentration of carbon dioxide (CO2) and other greenhouse gases, which can be substantially linked to the actions of humans ever since the beginning of the Industrial Revolution, is the fundamental factor that has led to the current state of global warming.

The Consumption of Fossil Fuels:

The burning of fossil fuels like coal, oil, and natural gas for the generation of energy and the conveyance of that energy is the primary contributor to the emission of carbon dioxide (CO2). The widespread reliance on these carbon-intensive energy sources has resulted in a huge increase in atmospheric CO2 levels, which in turn has led to serious repercussions for the global climate system.

Changes in Land Use and Clearing of Forests:

Deforestation, which is mostly caused by agricultural expansion, urbanization, and the exploitation of lumber, is a contributor to global warming because it lowers the capacity of the earth to absorb carbon dioxide through photosynthesis. The cutting down of trees not only reduces the amount of natural carbon sinks on the earth, but it also causes previously sequestered carbon to be released into the atmosphere, which makes the greenhouse effect even worse.

Processes Within Industry and Their Emissions:

Several different chemical processes and operations that require a lot of energy are responsible for the emission of greenhouse gases that are produced by industrial activities such as manufacturing, mining, and the production of cement. Industrial emissions of carbon dioxide (CO2), methane (CH4), and other pollutants make a considerable contribution to the overall burden of greenhouse gases, which causes the earth to warm up and alters the patterns of the global climate.

Agricultural Methods and the Emissions Caused by Livestock:

Agriculture, and more specifically contemporary methods of intensive farming, is one of the major contributors to the phenomenon of global warming. Both the keeping of cattle, which results in the production of methane, and the usage of synthetic fertilizers, which results in the release of nitrous oxide (N2O), are substantial contributors to the emissions of greenhouse gases produced by the agricultural sector.

In addition, the growing of rice in flooded paddies can result in the production of substantial quantities of methane, which is another factor that contributes to the warming of the planet.

Additional Contributing Factors to the Warming of the Planet

In addition to the direct impact that human activities have, there are several additional elements that can alter the dynamics of global warming. Natural occurrences, such as volcanic eruptions and shifts in the sun's radiation, have the potential to have momentary effects on the climate of the Earth. On the other hand, these natural factors typically have a short lifespan and cannot explain for the persistent warming trend that has been observed over the past few decades.

Loops of Feedback and Effects That Are Amplified:

The existence of feedback loops that can speed up the warming process is one of the troubling features of global warming. For instance, the melting of polar ice caps causes a reduction in the albedo of the Earth, which leads to an increase in the amount of solar radiation that is absorbed, which further accelerates the warming of the planet. In a similar vein, the thawing of permafrost causes the release of stored methane and carbon dioxide, which contributes to an increase in emissions of greenhouse gases.

Efforts and Agreements Made at the International Level:

In light of the fact that this is a global problem, efforts have been made on a global scale to investigate and solve the factors that contribute to it as well as its effects. The United Nations Framework Convention on Climate Change (UNFCCC) is a framework for international collaboration with the goal of stabilizing concentrations of greenhouse gases in the atmosphere at a level that prevents hazardous human interference with the climate system. The convention was established by the United Nations in 1992.

Strategies for Mitigation as well as Adaptation:

At the national and international levels, numerous policies have been developed and put into action in an effort to minimize both the causes and the impacts of global warming. The transition to renewable energy sources, improvements in energy efficiency, the implementation of sustainable land use practices, and the enhancement of carbon sequestration through reforestation and afforestation operations are all included in these measures. In addition, adaption measures, such as the construction of infrastructure that is resistant to the effects of climate change and the implementation of early warning systems for extreme weather events, are essential to the process of reducing the negative effects of global warming on communities and ecosystems.

The problem of global warming and the factors that contribute to it are intricate and multi-faceted, and they call for joint action on a worldwide scale.

Societies can work toward implementing sustainable practices and policies that mitigate the effects of climate change and safeguard the well-being of current and future generations if they understand the scientific principles that underlie global warming and acknowledge the human activities that are driving its progression. This can be accomplished by having an understanding of the scientific principles that underlie global warming and by acknowledging the human activities that are driving its progression. It is absolutely necessary for people, communities, governments, and international organizations to work together to address the underlying causes of global warming and to promote a future that is more sustainable and resilient for everyone.

2. **The importance of understanding the effects on melting ice and rising seas**
The climate of the Earth is altering at a rate that has never been seen before, with the melting of polar ice and the resultant increase in sea levels being one of the most apparent and significant symptoms of this change. The fragile equilibrium of the world's cryosphere, which consists of the polar ice caps, glaciers, and permafrost, is being thrown off by the warming of the planet. These changes will have a significant and far-reaching impact on ecosystems, coastal communities, and patterns of climate all around the world. It is not merely a matter of scientific interest to understand the effects of melting ice and rising oceans; rather, it is a matter of utmost relevance for the present and future of our planet.

1. The Effect on the Ecosystem:
The effect that melting ice and rising sea levels have on ecosystems is one of the most significant repercussions of climate change. As a result of the melting of the polar ice caps, certain regions, in particular, are going through severe changes. For instance, the Arctic, which is extremely sensitive to even minute shifts in temperature, is currently seeing the melting of sea ice, which is the natural habitat of a great number of different species, such as polar bears and seals. These animals will have to travel further distances in order to locate food when the ice melts, which may result in a reduction in their chances of survival as well as a change in their migration habits.

In addition, the melting of ice in the polar areas is having an effect on the distribution of marine life as well as its behavior. Fish species are moving into new habitats, which is having an effect on local fishing sectors. Alterations in the sea ice cover can also cause disruptions in the food chain by having an effect on the availability of phytoplankton, which is an essential source of nutrition for a wide variety of marine animals.

2. Vulnerability of Coastal Areas:
Coastal communities everywhere in the world are facing a serious risk as a result of rising sea levels.

It is becoming more likely that low-lying coastal areas may be flooded as a result of the melting of ice from polar regions and glaciers, which contributes to the overall swelling of the oceans. Homes, infrastructure, and people's livelihoods are all put in jeopardy as a result of the increasing frequency and severity of coastal erosion, storm surges, and flooding.

Rising sea levels present a challenge for coastal communities of all sizes, including the world's most populous cities, such as New York, Tokyo, and Mumbai, as well as tiny coastal towns. The rise in sea levels could force the relocation of millions of people, which would wreak havoc on the economy and exacerbate an already precarious humanitarian situation.

3. Climate Patterns Around the World:

The territories that are directly impacted are not the only places that will feel the effects of ice melting and rising sea levels. Alterations in the cryosphere have the potential to have far-reaching effects on the patterns of global climate. For instance, the loss of sea ice in the Arctic causes a disruption in the polar vortex, which is a wind pattern that circulates around the Arctic and has an effect on the weather across the Northern Hemisphere. This can lead to weather events that are more severe and unpredictable, which can have repercussions for agriculture, water supplies, and economies all around the world.

The release of methane, a powerful greenhouse gas, from thawing permafrost is another factor that contributes to global warming. This further complicates efforts to reduce the effects of climate change. Because of the interrelated nature of the Earth's climate system, changes in one location can have ripple effects that spread to other areas, resulting in a complicated web of cause and effect.

4. Resources Relating to Freshwater:

The retreat of glaciers has a direct and negative effect on the availability of freshwater resources, particularly in mountainous locations. Glaciers provide the function of natural reservoirs by holding water in the form of ice and gradually releasing it. This process is essential for ensuring that people located further downstream have a consistent supply of freshwater. This essential water source is in danger as glaciers melt and recede because of climate change.

A significant number of people obtain their drinking water, irrigate their land, and generate hydroelectric power from rivers that are supplied by glaciers. Especially in areas of the world where glaciers are the dominant source of freshwater, the depletion of this resource can result in severe water shortages, difficulties in agricultural production, and energy shortfalls.

5. Conservation of Biodiversity and Its Resources:

The effects of melting glaciers and rising sea levels can also be seen in the context of biodiversity and efforts to preserve it. As a result of the rapid changes taking place in their habitats, a great number of species that have developed specialized adaptations that allow them to survive in polar and glacial conditions are in danger of becoming extinct. However, the rate of change frequently outpaces the efforts that conservationists and scientists are making to document and safeguard these species. They are racing against the clock.

In addition, changes in sea levels and warming oceans have an effect on coral reefs, which are particularly susceptible to variations in temperature. Coral bleaching, which is brought on by rising ocean temperatures, poses a severe risk to these varied ecosystems. Because coral reefs provide a home to a diverse population of marine species and are necessary to the functioning of fisheries, it is critical that they be protected in order to maintain the health of both the marine environment and the human population.

6. The implications for society and the economy

The melting of ice and the subsequent rise in sea level will have repercussions not just on the environment but also on society and the economy. Adapting to and mitigating the effects of rising sea levels can place significant financial burdens on coastal towns. The construction of seawalls and flood defenses, as well as the relocation of infrastructure and towns, are included in these expenses.

Additionally, the real estate, insurance, agricultural, and tourism industries are the most susceptible to the effects of these changes in the economic landscape. The increased likelihood of flooding can have a negative impact on property prices in coastal locations, and the cost of insurance premiums may become un-affordable for residents. Saltwater intrusion may result in decreased agricultural productivity, particularly in low-lying areas of the world. Storms that are both more frequent and severe, as well as erosion, can have a significant negative influence on the tourism business in coastal regions.

7. Protection of the Country:

The ramifications of melting glaciers and rising seas for national security are equally significant. As sea levels continue to rise and more severe weather events occur on a regular basis, states will need to take precautions against the possible security risks posed by displaced populations, dwindling resources, and increased competition for arable land. The effects of climate change, which are worsened by the melting of ice and the rising of sea levels, can lead to resource conflicts and create a need for both humanitarian and military solutions.

8. The Relationship Between Indigenous Communities and Cultural Heritage:

The effects of melting ice can also be dangerous to cultural heritage and communities of indigenous people. Archaeological sites in polar regions are susceptible to harm from changes in the ice and the permafrost, which puts priceless cultural artifacts and traditional knowledge in jeopardy. Because their traditional methods of life are in danger, indigenous people who rely on ice-based activities like hunting and fishing risk cultural disruptions as well as economic hardships.

9. A Problem on a Global Scale:

Because global warming is a multifaceted, interrelated, and worldwide problem, having a good understanding of the implications of things like melting glaciers and increasing sea levels is essential. The repercussions are not limited to a single geographical area but rather influence all parts of our interconnected planet. In order to find a solution to this problem, worldwide collaboration and an integrated strategy that takes into account the problem on all of its levels—scientific, economic, social, and environmental—are required.

10. A Chorus of Calls to Action:

It is not sufficient to simply acknowledge the significance of the task of comprehending the implications of the melting ice and rising oceans; action is required. It is absolutely necessary to make measures to mitigate climate change by lowering emissions of greenhouse gases, shifting to more sustainable energy sources, and protecting natural carbon sinks like forests. Equally as significant is the implementation of adaptation techniques, such as the building of infrastructure that is resistant to damage and the creation of early warning systems.

In addition, it is necessary to collaborate internationally and factor in the effects of climate change when formulating public policies and taking other important decisions. It is imperative that we take immediate action to address the underlying causes of global warming and adopt sustainable habits that can assist in reducing the effects of glaciers melting and sea levels rising. To protect the world and assure a sustainable future for future generations, we face a problem that calls for individuals, communities, governments, and international organizations to come together, demonstrate their commitment, and work together.

3. **The structure and purpose of the book**

Books are vessels that transport readers on intellectual adventures, guiding them through information and insights that have been carefully curated. Books are not just warehouses of knowledge; rather, they are vessels that transport readers on intellectual excursions. The organization and goal of a book are two of the most important factors that influence the experience that a reader has while reading the book, as well as the way that material is presented and the framework that is provided for comprehending the book's primary concepts and ideas. During this investigation, we will delve into the relevance of the structure and purpose of a book, utilizing it as a lens to comprehend how the story progresses and the message it aims to communicate.

I. Opening Remarks and Establishing the Scene

The introduction acts as a gateway into the rest of the book, providing a thorough review of the core issue, the context within which it is set, and the goals that the author hopes to accomplish. It provides a road map for the reader, describing the primary issues, concerns, or questions that will be discussed throughout the work. The introduction establishes the mood for the remainder of the reading experience by picking the interest of the reader and creating a sense of anticipation for what is to come in the rest of the text.

II. The Organization of the Chapters and Their Flow: A Logic-Based Progression

The organization of a book often consists of a series of chapters that are organized in a sequential fashion, each one expanding upon the information presented in the one that came before it to produce a continuous storyline. Depending on the nature of the subject matter, the chapter arrangement may be ordered chronologically, thematically, or on the basis of a cause-and-effect relationship. These three organizational methods are all possible. A book that is well-structured will ensure that the flow of information is consistent and will lead the reader through a carefully planned sequence of concepts that will progressively increase their level of comprehension of the subject matter.

III. Section Headings and Navigational Aids: Getting Around in the Content

The use of subheadings and other forms of signposting within chapters can be thought of as navigational aids since they draw the attention of the reader to particular topics or areas of the text. They give a visual structure that breaks down difficult information into manageable chunks, which makes it easier for users to navigate through the text as a whole. Subheadings make it easier for readers to recognize important points, concepts, and arguments, allowing them to connect with the information in a manner that is more concentrated and well-organized.

IV. Aims and Objectives: Passing Along the Message

Every book is written with the intention of conveying some kind of message, argument, or point of view to its audience in some way, and every book has a purpose. The aim of the book is what drives the structure, which in turn shapes the material and guides the author in the choices he makes for the narrative. Whether the goal is to instruct, motivate, entertain, or stimulate critical thought, it acts as the guiding principle that determines the general direction and substance of the book.

V. Evidence and examples to back up the argument

When writing a book, authors frequently use illustrative material, such as supporting evidence, examples, case studies, and anecdotes, to drive home the point they are trying to make and the main message of the work. These components not only lend legitimacy and depth to the information presented in the book, but they also offer the audience real-world background and experiences that are similar to their own, which helps to highlight the author's most important points. The reader's comprehension is enhanced, and a deeper connection is fostered to the book's key topics, by the use of supporting data and examples that have been carefully selected.

VI. A Consolidation of Thoughts and Observations

The final chapter of the book acts as a synthesis of the various ideas, arguments, and realizations that have been given throughout the course of the tale. It restates the aim of the book, provides a summary of the most important points, and emphasizes how significant the knowledge that was provided is. In many cases, the conclusion will invite readers to reflect on the implications of the content of the book, urging them to examine how the knowledge learned may be applied in their own lives or in situations that are more general.

VII. Encouraging Exploration with the Provision of a Bibliography and Additional Reading

The inclusion of a bibliography or a list of suggested additional reading gives the reader the option to investigate similar topics or to go more deeply into certain areas of interest. These resources provide a channel for readers to deepen their knowledge beyond the boundaries of the book, giving them the ability to continue their intellectual research and engage with a variety of perspectives and scholarly works.

VIII. A Call to Action: Motivating Transformation and Participation

Some books, particularly those that focus on social issues, self-improvement, or advocacy, may include a call to action that urges readers to apply the knowledge obtained and take significant steps toward personal or societal change. This type of call to action is more common in books that focus on social issues, self-improvement, or advocacy.

The "call to action" section may provide actionable suggestions, tools, or techniques that enable readers to become active participants in the process of bringing about positive change or solving particular difficulties that were covered in the book.

IX. Contextualizing the Content within the Foreword and the Preface

A foreword or prelude may be included in certain publications in order to supply the reader with extra context, background information, or insights into the author's aims and motivations. The reader is given a look into the origins of the book, the author's personal relationship to the subject matter, or the larger context within which the work is located by reading these opening portions.

X. Acknowledgments: An Expression of Gratitude for People's Contributions

In the section titled "acknowledgments," the author has the opportunity to convey feelings of gratitude and appreciation to individuals and organizations that have played a part in the production of the book and its subsequent publication. It is a method to honor the collaborative efforts, support, and guidance that have molded the book's creation and publication journey. Specifically, it is a way to acknowledge those who have contributed to the book.

Chapter 1

Understanding Global Warming

In the most recent decades, there has been a substantial increase in awareness regarding the urgent environmental problem of global warming. It refers to the gradual rise, over a period of time, in the average surface temperature of the Earth as a result of both natural and human-caused factors. It is absolutely necessary to have a comprehensive understanding of the complexity of global warming, including its causes, effects, and potential remedies, in order to develop successful methods to address and mitigate the repercussions of this phenomenon.

The Science Behind a Changing Climate

The greenhouse effect is a natural occurrence that is responsible for the temperature on Earth remaining at a level that is suitable for human habitation. This effect is at the root of global warming. Greenhouse gases, such as carbon dioxide (CO_2), methane (CH_4), and water vapor, are responsible for this process, which entails the trapping of solar radiation in the atmosphere of the Earth by such greenhouse gases. However, human actions such as the burning of fossil fuels, the destruction of forests, and the processes involved in industrial production have considerably intensified the greenhouse effect, which has led to an imbalance in the climatic system of the Earth.

What's Causing All This Global Warming?

The burning of fossil fuels for the generation of energy and transportation is one of the key contributors to global warming because it causes the emission of significant quantities of carbon dioxide and other greenhouse gases into the atmosphere. Deforestation and other changes in land use also contribute to global warming because they reduce the capacity of the planet to absorb and store carbon. The manufacture of particular chemicals and the incorrect treatment of organic waste are two examples of industrial operations and waste management practices that further exacerbate the problem by producing harmful greenhouse gases. Other examples include the improper disposal of electronic trash.

The Repercussions of Climate Change

The effects of global warming are far-reaching and have an effect on many different aspects of the ecosystems that exist on Earth as well as human societies. These repercussions include changes in the frequency and severity of natural catastrophes, alterations in precipitation patterns, disturbances to biodiversity and ecosystems, and dangers to human health and well-being. The sea levels are also increasing as a result of these changes.

Some of the apparent symptoms of global warming, such as the melting of polar ice caps and glaciers, the acidification of the ocean, and the expansion of heatwaves and droughts, underline the urgency of tackling this issue.

Strategies for Problem Solving and Risk Management

To combat the effects of global warming, we need an approach that incorporates a variety of strategies, including international collaboration, government initiatives, technical advancements, and individual behavioral adjustments. The transition to renewable energy sources, the implementation of energy-efficient practices, the promotion of sustainable land management and afforestation, the enhancement of waste management systems, and advocacy for environmentally conscious policies and regulations at the international and national levels are all examples of strategies that fall under the category of "mitigation."

It is essential to get an understanding of global warming in order to cultivate a communal feeling of responsibility and the sense of urgency necessary to battle climate change. We can work toward reducing the effects of global warming and protecting the earth for future generations if we acknowledge the connection of human activities and the natural world and embrace policies and practices that are sustainable.

1.1 A comprehensive explanation of the greenhouse effect

The greenhouse effect is a fundamental and natural phenomenon that maintains the temperature of the Earth, allowing it to sustain circumstances that are favorable to the continued existence of life. However, human actions, in particular the burning of fossil fuels and the destruction of forests, have accelerated this process, which has led to both an increase in global temperature and a shift in the climate. In order to obtain an in-depth comprehension of the greenhouse effect, it is necessary for us to investigate the underlying principles of the phenomenon, as well as its significance for the climate of the Earth and its part in the continuous environmental difficulties we are currently facing.

A Natural Process That Causes the Greenhouse Effect

The phenomenon known as the greenhouse effect is a natural phenomenon that takes place when particular gases in the atmosphere of the Earth trap heat from the sun. The planet's surface is maintained at a temperature that is higher than it would be if these gases were not present because the heat that is trapped functions as a thermal blanket. Carbon dioxide (CO_2), methane (CH_4), water vapor (H_2O), and nitrous oxide (N_2O) are the principal greenhouse gases that are responsible for this occurrence. If the greenhouse effect did not exist, the average temperature of the Earth would be substantially lower, making the planet unfriendly to the majority of living things.

Harmonization of Radiant Forces

It is vital to take into consideration the radiative energy balance of the Earth in order to have a proper understanding of the greenhouse effect. The surface of the Earth is exposed to incoming solar radiation, the majority of which is in the form of visible light. This light is able to break through the Earth's atmosphere. The surface of the Earth then warms up as a result of the Earth's ability to absorb this solar energy. In a reversal of roles, the Earth is responsible for the emission of infrared radiation, which is a source of thermal energy.

Absorption and Release of Thermal Energy

The greenhouse gases that are present in the atmosphere are responsible for absorbing a portion of the infrared radiation that is emitted from the planet and preventing it from traveling directly into space. They instead reradiate the heat energy in all directions, including back towards the surface of the earth. Because of this process, the temperature of the surface is higher than it would be if the heat were simply permitted to escape in an unrestricted manner. In practice, the atmosphere performs the function of an insulator for heat, ensuring that the amount of energy that is taken in is equal to the amount that is given off.

Importance of Contributing to the Greenhouse Effect

The temperature of the earth must be kept within a certain range in order to sustain life as we know it, and the greenhouse effect plays a critical role in achieving this goal. Without it, the temperature of the earth would be too low to allow for the wide variety of ecosystems and climates that exist on it now. It is a natural regulatory mechanism that makes sure the temperature of the Earth's surface stays relatively the same, which in turn makes it possible for there to be liquid water and an environment that is habitable for a wide variety of species, including humans.

The Function of Vaporized Water

The abundance of water vapor, which is the most common greenhouse gas found in the atmosphere, is a substantial contributor to the phenomenon known as the greenhouse effect. Although the amount of water vapor present can change depending on the temperature as well as the location, the role that it plays in maintaining a consistent temperature is extremely important. When the atmosphere warms due to the presence of other greenhouse gases, water vapor functions as a feedback mechanism by allowing the atmosphere to hold more moisture, which in turn amplifies the warming impact.

Contextualization of the Past

Over the course of Earth's history, the greenhouse effect has been one of the most influential factors in determining the climate. Over the course of millions of years, this natural process has contributed to the preservation of a climate that is relatively stable, which has made it possible for life to evolve and adapt to changes in the environment.

This equilibrium has been thrown off by human activities, which have led to an increase in the concentration of greenhouse gases in the atmosphere. This, in turn, has contributed to a worsening of the greenhouse effect and an acceleration of global warming.

The amplified effect of the greenhouse gas

The natural greenhouse effect is necessary for sustaining a climate that is suitable for human habitation, but the augmented greenhouse effect that has resulted from the increased concentration of certain greenhouse gases as a result of human activity is needed. This amplified effect is driving the average temperature of the Earth to rise, which is leading to global warming and the problems that come along with it.

The Role of Human Activity and the Emission of Greenhouse Gases

A significant amount of greenhouse gas emissions come from the combustion of fossil fuels, such as coal, oil, and natural gas, which are burned for the generation of energy and the movement of goods. These activities result in the emission of greenhouse gases such as carbon dioxide (CO_2) and others into the atmosphere. In addition, the reduction of the planet's capacity to absorb and store carbon as a result of deforestation and other changes in land use is another factor that contributes to rising CO_2 levels.

Other Important Greenhouse Gases to Consider

Other greenhouse gases, in addition to carbon dioxide, play a key part in the intensification of the greenhouse effect. Methane (CH4) is released into the atmosphere by a variety of processes and activities, including as the digestion of food by cattle, the cultivation of rice, and the production and transportation of fossil fuels. Methane is a powerful greenhouse gas that is capable of retaining heat more effectively and for a shorter period of time than carbon dioxide does.

Nitrous oxide, also known as N2O, is another powerful greenhouse gas that is produced as a byproduct of agricultural and industrial processes, as well as the combustion of solid waste and fossil fuels. Nitrous oxide has a significantly higher ability to trap heat in the atmosphere than carbon dioxide does, and it can also remain there for a number of decades.

Modeling and Forecasting of the World's Climate

Scientists study the climate system of the Earth using global climate models (also known as GCMs) in order to gain a better understanding of the implications of the amplified greenhouse effect. In order to forecast potential shifts in patterns of temperature and precipitation in the future, these models take into account a wide range of elements, including as the concentrations of greenhouse gases, solar radiation, cloud cover, and ocean currents.

The forecast calls for an increase in temperature.

The greater greenhouse effect is expected to result in a wide range of potential temperature rises, which are projected by global climate models. They anticipate that the average temperature of the Earth could climb by several degrees Celsius by the end of the 21st century under a variety of emission scenarios. These increases in temperature could have significant repercussions for ecosystems, weather patterns, sea levels, and human societies.

Variability of the Climate Across Regions

The regional climate variability is another aspect that may be better understood thanks to climate models. In certain areas, temperatures could rise by a greater amount, while in others, changes in precipitation patterns, an increase in the frequency of heat waves, or an increase in the intensity of storms could occur. The research conducted on climate pays a lot of attention to the repercussions that global warming will have on certain regions.

The Effects That Are Caused By Climate Change

The acceleration of global warming, which is caused by an increase in the greenhouse effect, has a wide variety of effects, many of which are felt by the ecosystems and human societies of the planet. The development of methods to adapt to and lessen the effects of climate change requires a solid understanding of the repercussions that will be brought about by this phenomenon.

Increasingly High Sea Levels

The rise in sea levels is one of the most obvious and direct effects that can be seen as a direct result of global warming. The ice caps and glaciers in the polar regions will melt as the earth warms, and increasing temperatures will cause ocean to expand. These factors contribute to rising sea levels, which can lead to the erosion of coastal areas, an increase in the risk of floods, and the relocation of communities that are located in coastal areas.

Conditions of Extreme Weather

There has been a correlation between the rise in global temperature and an increase in the frequency and intensity of extreme weather occurrences. These occurrences include heatwaves that are more extreme, hurricanes, droughts, and heavy rainfall that are occurring more frequently. Extreme weather can have a wide range of implications, from the failure of crops and increased risk of hunger to the destruction of infrastructure and the loss of human life.

Ecosystems in Danger of Being Disturbed

The natural environment is extremely sensitive to variations in both temperature and climate. The disruption of ecosystems caused by global warming has an effect on the behavior and distribution of animals.

Some species may be threatened by the destruction of their habitat, an increase in competition for resources, or changes in their migration patterns. These disturbances can have a domino effect, having an adverse influence on entire ecosystems as well as the species inside them.

Acidification of the Ocean

Ocean acidification is a result of the oceans across the world absorbing more carbon dioxide than they need to. This phenomena has an effect on marine life, and more specifically on species that have calcium carbonate shells or skeletons, such corals and mollusks. The acidification of the ocean has the potential to have a domino effect on marine food webs as well as fisheries.

Consequences for Human Health

The health of humans can be adversely affected both directly and indirectly by global warming. It is anticipated that there would be a spike in the number of deaths and diseases caused by heat as temperatures continue to climb. Alterations in climate can also have an effect on the occurrence and distribution of diseases that are transmitted by vectors, such as dengue fever and malaria. These negative effects on health are more likely to be experienced by vulnerable groups, in particular those living in poor nations.

Strategies and Answers for Damage Mitigation

A concentrated effort is required on the local, national, and international levels in order to address the issue of global warming and reduce the impact of its repercussions. There are a number of different approaches and options available to cut down on emissions of greenhouse gases and to lessen the impact of global warming.

Make the Switch to Alternative Energy Sources

One of the most important things that can be done to mitigate the effects of climate change is to
switch from using fossil fuels to renewable energy sources like solar, wind, and hydropower. This change will minimize greenhouse gas emissions from energy production while also decreasing reliance on energy sources that produce a high amount of carbon.

Effective Use of Energy

Increasing energy efficiency across all sectors—transportation, buildings, and industries—leads to lower overall energy consumption as well as less emissions of greenhouse gases. This strategy calls for the implementation of procedures and technology that are more energy-efficient.

Both reforestation and afforestation are being done

Both reforestation, which is the process of replanting trees in regions that have been deforested, and afforestation, which is the process of creating new forests, contribute to the removal of carbon dioxide from the atmosphere. Because of these activities, the planet's capacity to absorb and store carbon is increased, which helps to mitigate the increased greenhouse impact.

Management of the Land That Is Sustainable

It is possible to lower emissions caused by deforestation and changes in land use through the implementation of practices such as sustainable agriculture and responsible land management. They advocate for the sustainable management of land, the preservation of natural areas, and the defense of ecosystems.

The Management of Waste

It is possible to lower emissions of methane and other greenhouse gases from landfills and waste treatment facilities by implementing better waste management techniques. Some of these practices include reducing the amount of garbage generated, recycling materials, and composting waste.

The Global Response to Climate Change and Related International Agreements

The creation of climate policies, legislation, and agreements at the national and international levels is critically important to the fight against global warming. Agreements such as the one that was reached in Paris establish goals for the reduction of emissions and create a framework for international collaboration on climate change action.

It is absolutely necessary to have a complete comprehension of the greenhouse effect in order to effectively address the issues that are brought by global warming. This natural process, which has been regulating the temperature of the Earth for millions of years, is being magnified by human activities, which is leading to a multiplicity of repercussions, including rising temperatures, extreme weather events, and disruptions to ecosystems. Human activities have enhanced this natural process, which has been regulating the temperature of the Earth for millions of years. In order to slow or stop the progression of global warming, we need to make a concentrated effort to cut emissions of greenhouse gases, convert to cleaner energy sources, and encourage sustainable activities. We have the ability, through concerted effort, to reduce the severity of the effects of global warming and ensure the survival of our planet and all of its unique ecosystems in the years to come.

1.2 Historical context and the discovery of global warming

The historical setting of global warming may be traced back to the 19th century, when scientists for the first time started investigating the connection between the composition of the atmosphere and the temperature of the earth. Our understanding of the greenhouse effect and the influence that human activities have on the climate of the Earth has been significantly advanced over the course of the years thanks to a number of important discoveries and improvements in scientific understanding. In the course of this investigation, we will look into the historical background as well as the seminal discoveries that led to the identification of global warming as an urgent environmental concern.

Investigations During the Formative Years of Science
The 19th century as a whole

Joseph Fourier and John Tyndall were among the pioneering researchers who worked on the energy balance of the Earth and the role that gases in the atmosphere play in maintaining a constant temperature at the beginning of the 19th century. In the year 1824, the French mathematician Fourier was the first person to introduce the idea of the greenhouse effect. He suggested that the atmosphere of the Earth functions as a blanket, retaining heat and ensuring a consistent climate that is favorable to the support of life.

The Hypothesis Concerning the Greenhouse Effect

John Tyndall, an Irish scientist, conducted experiments in the middle of the 19th century that established the heat-absorbing capabilities of some gases, particularly water vapor and carbon dioxide. These findings built on the work that Fourier had previously accomplished. The studies conducted by Tyndall helped to pave the way for a better understanding of the greenhouse effect and the part it plays in the climate system of the Earth.

Early Awareness of Warming in the Global Environment

The theory of carbon dioxide first proposed by Svante Arrhenius

Svante Arrhenius, a Swedish scientist, put up a hypothesis in a paper that was published in 1896 that suggested there was a correlation between the amount of carbon dioxide in the atmosphere and the temperature of the earth. The studies conducted by Arrhenius indicated that human activities, such as the burning of fossil fuels, could contribute to an increase in atmospheric CO_2 levels, which would eventually result in a gradual warming of the Earth's surface. His work created the framework for understanding the potential consequences of artificial greenhouse gas emissions. Initially, Arrhenius saw this warming as a positive development that could avert the beginning of a new ice age. However, his work laid the groundwork for grasping these potential ramifications.

Transformations and Awakenings of the 20th Century

Increase in the Number of Efforts Put Into Research

Throughout the course of the 20th century, scientists continued their research into the intricate workings of the Earth's climate system and the variables that affect average temperatures around the world. Research institutes and organizations invested a substantial amount of resources to the research of atmospheric composition, oceanic currents, and climate patterns, which resulted in a clearer knowledge of the mechanisms driving global climate change.

The Origin and Development of Climate Models

The development of complex climate models that were able to replicate the functioning of the Earth's climate system was made possible by advances in computing technology. With the help of these models, scientists were able to investigate a wide range of climate scenarios and estimate the potential effects that human activities could have on the temperature and weather patterns of the world.

The growing awareness of, and concern for, environmental issues Increasing Public Consciousness Along with Growing Environmental Movements

In the latter half of the 20th century, an increase in public awareness of environmental issues, such as pollution in the air and water, deforestation, and the conservation of animals, brought the topic of global warming to the forefront of the discourse that was taking place on an international level. The importance of addressing the anthropogenic activities that are contributing to climate change has been more widely aware as a result of environmental movements and advocacy initiatives.

Consensus in the Scientific Community and Cooperative Efforts Internationally

The formation of the Intergovernmental Panel on Climate Change (IPCC) in the year 1988 was a key step forward in the fight against the effects of climate change on a worldwide scale. In order to evaluate the scientific evidence and propose suggestions for limiting the effects of global warming, the Intergovernmental Panel on Climate Change (IPCC) gathered together scientists, politicians, and other professionals from all over the world.

The Current State of Understanding and the Ongoing Research The Paris Agreement and the International Response to Climate Change

In 2015, the international community reached a significant milestone in its fight against climate change when it officially ratified the Paris Agreement. The goal of the agreement was to keep the increase in global temperature well below 2 degrees Celsius above pre-industrial levels, with the objective of pursuing measures to keep the increase in temperature to 1.5 degrees Celsius. The governments that signed the document made a commitment to lessen the negative effects of global warming by enacting climate action plans and reducing their emissions of greenhouse gases.

Recent Progress in the Field of Climate Science

Our comprehension of global warming and the myriad ways in which it manifests itself has been significantly improved by recent breakthroughs in climate research. These breakthroughs include satellite technology, sophisticated monitoring systems, and exhaustive data analysis. In spite of the fact that research on climate change is ongoing, ongoing attempts are still being made to refine climate models, assess the vulnerability of ecosystems and communities, and develop strategies for adaptation and resilience in the face of a changing climate.

When it comes to tackling one of the most critical problems of our era, the importance of scientific investigation and joint efforts cannot be overstated. Both the historical setting and the discovery of global warming highlight this significance. The path toward realizing the impact that human activities have on the climate of the Earth has been defined by important discoveries, awareness-raising initiatives, and international cooperation. This trip began with the early research of the greenhouse effect and has culminated in the modern understanding of climate science. Along the way, there have been critical discoveries. As we work through the intricacies of climate change, it is absolutely necessary that we continue to create a shared commitment to sustainability, resilience, and the preservation of our planet for the generations that will come after us.

1.3 Key contributors to greenhouse gas emissions

Emissions of greenhouse gases are the primary contributor to the current state of the climate problem. These emissions, which are mostly caused by human activity, serve as a blanket over the Earth's atmosphere, causing temperatures to rise along with a plethora of environmental, social, and economic concerns. It is vital to have an understanding of the primary contributors to greenhouse gas emissions in order to effectively address this serious problem. Additionally, it is important to investigate the sources and industries that play a large role in driving this worldwide challenge. This all-encompassing investigation will provide a full summary of these environmental elements and the impact they have on the environment.

Before we begin:

Greenhouse gases, often known as GHGs, are gases that are found in the atmosphere of the Earth and are responsible for preventing heat from escaping into space. This natural occurrence, which is referred to as the greenhouse effect, is essential for keeping the temperature of the planet within a range that allows for the continued existence of life. However, the concentration of greenhouse gases in the atmosphere has been significantly increased due to human activities such as the burning of fossil fuels, the destruction of forests, and the operation of industrial processes. This has resulted in a stronger greenhouse effect.

Carbon dioxide (CO2), methane (CH4), nitrous oxide (N2O), and fluorinated gases are some of the most important greenhouse gases. These gases have different capacity for retaining heat and different lives in the atmosphere, both of which make them substantial contributors to both global warming and climate change.

Carbon Dioxide, often known as CO2

The Main Contributor to the Greenhouse Effect

Carbon dioxide, sometimes known as CO2, is the most common and well-known greenhouse gas. It is also the primary contributor to the warming of the planet that is caused by humans. It is largely produced as a byproduct of the combustion of fossil fuels, including as coal, oil, and natural gas, which are used in the generation of electricity and the movement of goods. In addition, the reduction of the planet's capacity to absorb and store carbon due to changes in land use and deforestation results in the emission of carbon dioxide into the atmosphere.

Principal Carbon Dioxide Emissions Sources

Production of Energy: A significant contributor to greenhouse gas emissions is the combustion of fossil fuels for the generation of electricity and heat in power plants. The combustion of natural gas and coal, as well as the processing of petroleum, all contribute significantly to the emission of carbon dioxide into the atmosphere.

Transportation: The transportation industry on a worldwide scale, which includes automobiles, trucks, airplanes, and ships, is a substantial contributor to the emission of carbon dioxide (CO2). A significant amount of pollution is produced as a result of our reliance on internal combustion engines, in particular those that are fueled by gasoline and diesel.

Industrial Processes: A wide variety of industrial procedures, such as the production of cement and steel and the processing of chemicals, result in the emission of CO2 as a waste product.

These emissions are frequently the result of the consumption of fossil fuels for the purpose of obtaining energy and heat.

Changes in land use, notably deforestation, result in the release of carbon that has been stored in trees and soil into the atmosphere in the form of carbon dioxide (CO2). The capacity of the globe to absorb CO2 through the process of photosynthesis is being hampered by the destruction of forest land.

Agriculture: The rearing of cattle and the usage of synthetic fertilizers are two examples of agricultural practices that contribute to the generation of carbon dioxide. Both the high energy requirements of contemporary agriculture and the methane emissions caused by cattle are major contributors to the problem.

Methane, or CH4

An extremely important greenhouse gas

Methane, also known as CH4, is a powerful greenhouse gas that is capable of retaining heat more effectively and for a shorter period of time than carbon dioxide does. It is emitted as a byproduct of multiple processes, some of which are natural and some of which are anthropogenic. Emissions of methane are a substantial contributor to the warming of the planet.

The Most Important Causes of Methane Emissions

Methane is produced during the digestive process of livestock such as cattle and sheep through a process called enteric fermentation. This enteric fermentation is responsible for a significant amount of total methane emissions and is intimately connected to the livestock business.

Rice Cultivation: When rice paddies are flooded, anaerobic conditions are created that are excellent for methane-producing microbes. This results in the emission of methane throughout the rice cultivation process.

Wetlands: Because they provide the right circumstances for the growth of microorganisms that produce methane, natural wetlands are a substantial source of methane emissions.

Production of Natural Gas and Oil: Leaks of the greenhouse gas methane can occur during the extraction and transport of natural gas and oil, which contributes to overall emissions.

Methane is produced when organic waste is decomposed in anaerobic circumstances, such as those found in landfills and wastewater treatment facilities, and this gas is released into the atmosphere.

Nitrous oxide, also known as N2O.

A Gas that Is Both Powerful and Persistent

Nitrous oxide, often known as N2O, is a powerful greenhouse gas that has a far higher ability to trap heat than carbon dioxide (CO2) and a longer atmospheric lifetime. It is emitted from a variety of sources, both naturally occurring and those driven by humans.

Nitrous oxide emissions come from a variety of primary sources

Agriculture: One of the biggest contributors to the emission of nitrous oxide is the use of synthetic fertilizers in agricultural production. Nitrous oxide is produced when microorganisms in the soil metabolize the nitrogen that comes from fertilizers.

Burning of Biomass: Nitrous oxide is released into the atmosphere when biomass, such as crop leftovers and wood, is burned.

Nitrous oxide emissions can be produced during the process of treating sewage and wastewater, which is referred to as "wastewater treatment."

Processes that take place in factories are one of the primary contributors to the emission of nitrous oxide. One example of this is the manufacturing of nitric acid.

Gases that contain fluorine

Fluorinated gases are synthetic substances that are utilized in a variety of applications within the manufacturing sector. These gases include hydrofluorocarbons (HFCs), perfluorocarbons (PFCs), and sulfur hexafluoride (SF6). Although they account for only a relatively insignificant part of total greenhouse gas emissions, the warming potential of each individual molecule is quite high.

Fluorinated gas emissions come mostly from the following sources:

Refrigeration & Air Conditioning: HFCs are a typical component of both refrigerant and refrigerant-free air conditioning and refrigeration systems. The manufacturing, utilization, and disposal of these systems all contribute to the emissions that they cause.

Insulating Gas in High-Voltage Equipment: SF6 Is utilized in the Electrical Industry SF6 is utilized as an insulating gas in high-voltage equipment in the electrical industry. Leaks and improper maintenance of equipment are the primary causes of emissions.

Production of Semiconductors: PFCs are essential components in the manufacturing of semiconductors. The manufacturing process results in the release of emissions.

Climate Pollutants With a Short Half-Life

In addition to the major greenhouse gases, short-lived climate pollutants, also referred to as SLCPs, play an important part in the process of global warming. These pollutants, which include black carbon (also known as soot), methane, and hydrofluorocarbons (HFCs), are responsible for a large amount of heat and can have rapid effects on climate change. Taking action to address SLCPs is a productive technique for reducing warming in the near future.

Important SLCP Emissions-Causing Sources

Cookstoves and Open Fires: Black carbon and methane are released into the atmosphere when biomass, such as wood and agricultural leftovers, is burned in cookstoves and open fires.

In the transportation industry, diesel engines in particular constitute a substantial contributor to black carbon emissions, whereas HFCs are utilized in the air conditioning systems of motor vehicles.

Agriculture is a major contributor to both methane and nitrous oxide emissions due to the management of livestock manure, which includes enteric fermentation. Additionally, the use of synthetic fertilizers also adds to these emissions.

Management of rubbish Both the open burning of rubbish in the open air and the management of landfills can lead to the release of methane and black carbon.

The Importance of Alterations in Land Use and Land Cover

Changes in land use and land cover, in particular deforestation and afforestation, play a significant impact in the production of greenhouse gases. Changes in carbon storage and emissions are contributed to by the conversion of natural landscapes into metropolitan centers, agricultural fields, and other land uses.

Destruction of Forests

The release of carbon that has been stored in trees and soil is a consequence of the removal of forests for agricultural and urban development and infrastructure purposes. This carbon is released into the atmosphere as carbon dioxide, which contributes to the warming of the planet.

Both initial forestation and subsequent reforestation

On the other hand, afforestation (the development of forests on land that was not previously wooded) and reforestation (the replanting of trees in previously deforested regions) help absorb carbon from the atmosphere, thereby limiting the impacts of global warming. Afforestation is the establishment of forests on land that was not previously forested.

Important Businesses and Activities Contributing to Emissions of Greenhouse Gases

In order to effectively create mitigation plans and regulations, it is essential to have a solid understanding of the industries and activities that are responsible for greenhouse gas emissions.

Vitality and vigor

The burning of fossil fuels to generate electricity and heat makes the energy industry a significant source of greenhouse gas emissions. These emissions are primarily caused by the usage of fossil fuels. The largest contributors of carbon dioxide emissions are power plants that burn coal, natural gas, and oil for fuel, while the transportation sector is highly reliant on gasoline and diesel fuel.

The agricultural industry

There are many different aspects of agriculture that contribute to the generation of greenhouse gases. Methane emissions are produced as a result of enteric fermentation in animals, nitrous oxide emissions are produced when synthetic fertilizers are used in crop cultivation, and energy-intensive contemporary agriculture is dependent on fossil fuels.

Alterations to Both Land Use and Land Cover

The release of carbon that had been stored in trees and soil, primarily as carbon dioxide, is caused by deforestation and other changes in land use. However, these emissions can be mitigated through activities such as afforestation and replanting.

Procedures in Manufacturing

The production of some types of industrial goods results in the emission of greenhouse gases. Manufacturing processes such as cement production, steel manufacturing, and chemical production are included in this category.

The Management of Waste

Methane is a strong greenhouse gas that can be released into the atmosphere as a result of inefficient waste management techniques such as open burning and poorly managed landfills.

Carriage and conveyance

The transportation industry, which encompasses road, rail, air, and sea transport, is dependent on internal combustion engines, which produce considerable volumes of carbon dioxide. These engines may be found in all modes of transportation. In addition, the usage of refrigerants in automobiles, especially HFCs, contributes to the production of emissions.

Structures that are both residential and commercial in nature

The consumption of energy by business and residential buildings for the purposes of heating, cooling, and generating electricity is a substantial contributor to the production of greenhouse gases. It involves the burning of fossil fuels for the purpose of producing both heat and power.

Strategies and Answers for Damage Mitigation

In order to counteract climate change and lessen its effects, addressing emissions of greenhouse gases is of the utmost importance. The transition to clean energy sources, improvements in energy efficiency, the promotion of sustainable land management, and the adoption of responsible waste management techniques are all effective solutions for mitigating climate change. In addition, a framework for global collaboration in the reduction of emissions is provided by rules, legislation, and international agreements such as the Paris Agreement.

The emissions of greenhouse gases, which are caused by human activities, have a significant impact on the climate system of the Earth. Together with other climatic pollutants that have a short half-life, carbon dioxide, methane, nitrous oxide, and fluorinated gases are all factors that contribute to climate change and global warming. The first step in developing viable solutions to reduce the consequences of climate change and protect the world for future generations is to have an understanding of the sources and industries that are accountable for these emissions. To address the issue of greenhouse gas emissions, there needs to be a concerted effort on a worldwide scale, involving governments, corporations, and individuals, to transition to practices that are sustainable and low-carbon, which can help lessen the effects of global warming.

| 34 |

Chapter 2

The Earth's Cryosphere

All of the frozen regions of our world make up what is known as the cryosphere, which plays an important role in the climate system of our planet. Glaciers, ice caps, sea ice, permafrost, and seasonal snow cover are all components of this feature. These icy components play a large part in the process of controlling the climate of the Earth, and the dynamics of their interactions have profound implications for both the natural world and human societies. In the course of this in-depth investigation, we are going to look into the cryosphere, its many different components, the relevance of those components, and the effect that climate change has on frozen places.

Before we begin:

The word "cryosphere" comes from the Greek words "kruos" (frost) and "sphaira" (sphere), and it refers to the region of the Earth's surface that contains water in a solid state. The phrase derives from the word "kruos," which means "frost." The cryosphere encompasses a diverse range of ecosystems, including the ice caps of the poles, glaciers in the mountains, and frozen tundra. Although it may appear remote and unimportant, the cryosphere actually plays a significant part in the process of controlling the climate of the Earth and has a direct influence on both the natural world and the activities of humans.

Structures that make up the Cryosphere

Ice Sheets, or Glaciers and Ice Caps

Large accumulations of ice that are present throughout the year are known as glaciers and ice caps. They originate as snow accumulates and becomes compressed over a period of time. Glaciers are widespread over the globe, and can be found anywhere from the arctic regions to the highest mountain ranges. Changes in glacier mass and extent reflect adjustments in temperature and precipitation patterns, hence changes in glacier mass and extent are important indicators of climate change.

Ice from the Sea

The freezing of ocean water results in the formation of sea ice, which blankets the polar oceans during the colder months of the year. It does this by reflecting sunlight back into space and acting as an insulator for the water below, both of which are essential functions in the process of maintaining a stable temperature on Earth.

The melting of Arctic sea ice, which is being caused by rising temperatures, will have significant repercussions for both the regional and global climate systems.

The permafrost

Permafrost is defined as soil or sediment that has a temperature that has remained at or below freezing (0 degrees Celsius or 32 degrees Fahrenheit) for at least two years in a row. High-latitude locations, such as the Arctic and the subarctic, are where it can be found the most frequently. The permafrost serves as a reservoir for carbon by storing the organic matter that has accumulated over a period of thousands of years. The melting of permafrost causes an increase in the amount of greenhouse gases, mainly methane, that are released into the atmosphere, which in turn exacerbates climate change.

A Snowfall That Occurs Seasonally

The snowfall that occurs during the winter months and then melts away throughout the spring or summer is considered seasonal snow cover. This particular variety of snow is common in both temperate and polar climates, and its presence has consequences for the water resources, temperature, and ecosystems of the region. Temperature and precipitation patterns have an effect on the length of time that seasonal snow cover lasts as well as its overall extent.

The icebergs

Icebergs are big chunks of ice that float freely in the water and are formed when glaciers and ice shelves lose chunks of ice and send them into the ocean. They are essential markers of glacial movements and can be hazardous to navigation, particularly in polar locations. Alterations in the stability of glaciers can be connected to alterations in iceberg calving patterns.

The Function of the Cryosphere in the Control and Modulation of Climate

The effect of albedo

The albedo effect is one of the most important ways in which the cryosphere has an effect on the climate. Ice and snow have a high reflectivity, which means that they reflect a considerable percentage of the solar radiation that is coming from the sun back into space. This property of being reflecting contributes to the preservation of frigid temperatures in polar regions and helps to keep the overall temperature of the planet stable. The ice-albedo feedback is a self-reinforcing loop that occurs when ice melts and is replaced by darker, heat-absorbing surfaces such as water or soil. This results in the albedo decreasing, which in turn leads to more warming and more ice melt.

Insulation from the heat

The presence of sea ice works as a thermal insulator, preventing heat from being transferred from the ocean to the atmosphere above it. It contributes to the stabilization of polar regions and helps to keep the temperature of the ocean underneath them constant. The melting of sea ice causes a disruption in this insulation, which in turn causes the oceans to warm up and alters the patterns of ocean circulation.

Carbon Sequestration

Large quantities of organic stuff, such as decomposed plant matter and carbon from long ago, can be found preserved in permafrost. When temperatures are below freezing, decomposition proceeds at a snail's pace, and there is only a small amount of carbon released in the form of methane and carbon dioxide. However, when temperatures rise enough to cause permafrost to melt, the carbon that has been stored there can be released into the atmosphere, which contributes to the production of greenhouse gases.

The Cryosphere's Reactions to the Effects of Climate Change

The Melting of Glaciers

There has been a noticeable acceleration in the retreat of glaciers all around the world. This retreat is a contributor to rising sea levels, which in turn affects communities and regions along the coast. In addition, glacier meltwater is a key source of freshwater for a large number of people, and the decline of this source can cause water scarcity in certain regions.

Decrease in Sea Ice

In particular, the Arctic is going through a period of fast reduction in the amount of sea ice cover. This change destabilizes the ecosystems of the surrounding area, puts the existence of arctic species in jeopardy, and has geopolitical repercussions because it paves the way for new shipping routes and increases access to resources.

Melting of the Permafrost

The melting of permafrost can cause ground to sink, which in turn can cause damage to buildings and other types of infrastructure. Additionally, the release of greenhouse gases caused by the thawing of permafrost contributes to the acceleration of global warming. This melting can also cause havoc for the means of subsistence and cultural practices that indigenous groups rely on.

Alterations to the Snow Cover

The shifting patterns of seasonal snow cover can have an effect on the amount of water that is available, in particular in areas where snowmelt is a key source of freshwater. Changes in the timing of the snowmelt can result in a lack of water, which will have repercussions for agriculture, ecosystems, and communities.

Consequences for Both the Region and the World as a Whole

The changes that are taking place in the cryosphere will have repercussions that are felt far beyond the polar regions. They have an effect on the patterns of weather, the rise in sea level around the world, as well as the severity and frequency of extreme weather occurrences. For instance, the reduction in the amount of sea ice in the Arctic can change the patterns of air circulation, which in turn can have an effect on the weather systems in places located in the middle latitudes.

The melting of glaciers and ice caps poses a threat to the water supply of millions of people who rely on meltwater for drinking, farming, and hydroelectric power. Coastal erosion and the flooding of low-lying coastal areas are both caused by increasing sea levels, which are caused by the melting of glaciers and ice caps. Additionally, the expansion of seawater as it heats also contributes to the problem.

Both Adaptation and Mitigation are Needed.

Adapting to the changes that are occurring in the cryosphere is a difficult challenge that requires reevaluating infrastructure, the management of water resources, and emergency preparedness measures. Communities in the Arctic, for instance, are formulating plans to combat the thawing of permafrost and the erosion of coastal areas as a result of climate change.

In order to mitigate the effects of changes in the cryosphere, it is necessary to reduce emissions of greenhouse gases in order to moderate the rate of warming. This is absolutely necessary in order to protect the cryosphere and prevent the occurrence of catastrophic effects. For the sake of the protection of both the cryosphere and the earth as a whole, it is imperative that global efforts be made to lessen the effects of climate change. This includes making the switch to renewable energy sources, cutting down on deforestation, and improving energy efficiency.

The cryosphere is an essential element of the climatic system of the Earth, and its state of health and stability has direct bearing on the state of the natural world, the ecosystems it supports, and human cultures. The cryosphere is undergoing major changes as a direct result of the ongoing rise in global temperatures caused by human activity. These changes include the melting of permafrost and the retreat of glaciers. These shifts have a wide range of repercussions, some of which include an increase in the level of the sea, disruptions in the patterns of weather, and dangers to freshwater resources.

It is absolutely necessary to have a solid understanding of the cryosphere and the function it plays in climate regulation in order to develop successful methods for adaptation and mitigation. In light of the challenges posed by a warming planet, it is of the utmost importance that we take concerted action to limit emissions of greenhouse gases, conserve the cryosphere, and devise strategies to adapt to the changes that are inevitably going to occur and are already in the process of happening. The protection of the cryosphere is not only an important issue for the planet's ecosystem, but it is also an essential factor in ensuring the planet's continued habitability over the long run.

2.1 The cryosphere defined and its significance in the global climate system

The term "cryosphere" refers to all of the frozen sections of our globe, including polar ice caps, mountain glaciers, permafrost, sea ice, and seasonal snow cover. The cryosphere is an essential part of the climate system that controls our planet's weather. This icy domain plays a large part in the process of regulating the climate of the Earth, which in turn has enormous repercussions for the planet's ecosystems, human societies, and the environment. During this in-depth investigation, we will define the cryosphere, investigate the numerous components that make up the cryosphere, and explain the fundamental significance that the cryosphere possesses in the complex web that is the global climate system.

Figuring Out What the Cryosphere Is

The word "cryosphere" comes from the Greek words "kruos" (frost) and "sphaira" (sphere), and it is used to refer to the region of the Earth's surface where water is found in a solidified, frozen condition. The term "cryosphere" refers to a varied range of settings that are distinguished by the presence of ice or snow. These environments include the vast icy plains of the polar regions, the high mountain ranges, and even the seasonal snowfall that occurs in the more temperate regions. The cryosphere is an essential component of Earth's climate dynamics, and it plays a critical role in the preservation of the fragile temperature equilibrium that exists on the planet.

Structures that make up the Cryosphere

1. **Ice Sheets and Continental Glaciers:**
   Glaciers and ice caps are both enormous accumulations of ice that are present throughout the year. They come into existence as a result of the slow accumulation and compression of snow over the course of time. Glaciers may be found on every continent, from the polar areas to the high mountain ranges, and they are an essential indicator of the changing climate. Glaciers can be found in both the polar regions and the high mountain ranges. Alterations in temperature and precipitation patterns are a reflection of changes seen in the mass and area of glaciers.

2. **Ice on the Sea:**
   The freezing of ocean water results in the formation of sea ice, which blankets the polar oceans throughout the winter months. It does this by reflecting a considerable fraction of the incoming solar radiation back into space, which in turn cools the globe. This is an essential part of the process by which the climate of the Earth is controlled. The melting of Arctic sea ice, which is being caused by rising temperatures, will have significant repercussions for climate systems both regional and global in scale.

3. **the term "permafrost":**
   Permafrost is defined as soil or silt that has been frozen for at least two years in a row, with temperatures remaining continuously below 0 degrees Celsius (32 degrees Fahrenheit). High-latitude

locations, such as the Arctic and the subarctic, are where it can be found the most frequently. The permafrost serves as a reservoir for carbon by storing the organic matter that has accumulated over a period of thousands of years. The melting of permafrost causes an increase in the amount of greenhouse gases, mainly methane, that are released into the atmosphere, which in turn exacerbates climate change.

4. **Snowfall that occurs seasonally:**
Snowfall that occurs just during the winter months and then melts away during the spring or summer is considered seasonal snow cover. It is common in both temperate and polar climates and plays an important part in the water supplies, climate, and ecosystems of the places in which it is found. Temperature and precipitation patterns have an effect on the length of time that seasonal snow cover lasts as well as its overall extent.

5. **Icebergs:**

Icebergs are big chunks of ice that float freely in the water and are formed when glaciers and ice shelves lose chunks of ice and send them into the ocean. They are vital indications of glacial movements as well as the potential for causing navigational problems, particularly in polar regions. Alterations in the stability of glaciers can be connected to alterations in iceberg calving patterns.

The Importance of the Cryosphere in Understanding the Global Climate System

1. **The effect of albedo:**
The albedo effect is one of the most important ways in which the cryosphere has an effect on the climate. Ice and snow are extremely reflecting surfaces, which reflect a significant amount of the solar radiation that is arriving from outside back into space. This high albedo contributes to the preservation of polar areas' frigid temperatures and the preservation of a stable global

temperature.

The albedo, on the other hand, will decrease as the ice melts and is replaced by darker surfaces that will absorb more heat, such as water or soil. This leads to an increase in temperature, an increase in the amount of ice that melts, and a cycle that reinforces itself called the ice-albedo feedback.

2. **Insulation from the Heat:**

Because it works as a thermal insulator, sea ice prevents heat from being transferred from the ocean to the surrounding atmosphere. It contributes to the stabilization of polar regions and helps to keep the temperature of the ocean underneath them constant. The melting of sea ice causes a disruption in this thermal insulation, which in turn causes the oceans to warm up and alters the patterns of ocean circulation.

3. **Carbon Sequestration:**

Large quantities of organic stuff, such as decomposed plant matter and carbon from long ago, can be found preserved in permafrost. When temperatures are below freezing, decomposition proceeds at a snail's pace, and there is only a small amount of carbon released in the form of methane and carbon dioxide. However, when temperatures rise enough to cause permafrost to melt, the carbon that has been stored there can be released into the atmosphere, which contributes to the production of greenhouse gases.

4. **The Dynamics of Glaciers**

Variations in both regional and global climate can be understood better with the use of information provided by changes in glacier mass and extent. The melting of glaciers is a contributing factor in rising sea levels, which in turn threatens the towns and regions that are located along coasts. In addition, glacial meltwater is a crucial source of freshwater for a large number of people, and a decrease in this source of water can result in a shortage of water in certain regions.

5. Controlling the Level of the Sea:

The cryosphere, in particular in the form of ice sheets and glaciers, is an important component in the process of controlling the levels of seawater around the world. The melting of ice adds to rising sea levels, which in turn has an effect on coastal areas and islands that are low-lying. It is absolutely necessary to have an understanding of these dynamics in order to accurately forecast and potentially mitigate future sea-level rise.

The Cryosphere's Reactions to the Effects of Climate Change
The Retreat of Glaciers:

There has been a noticeable acceleration in the retreat of glaciers all around the world. This retreat is a contributor to rising sea levels, which in turn affects communities and regions along the coast. In addition, glacial meltwater is a crucial source of freshwater for a large number of people, and the decline of this source can lead to a shortage of water in certain regions.

The Melting of Sea Ice:

In particular, the Arctic is going through a period of fast reduction in the amount of sea ice cover. This change destabilizes the ecosystems of the surrounding area, puts the existence of arctic species in jeopardy, and has geopolitical repercussions because it paves the way for new shipping routes and increases access to resources.

Thawing of the Permafrost:

The melting of permafrost can cause ground to sink, which in turn can cause damage to buildings and other types of infrastructure. Additionally, the release of greenhouse gases caused by the thawing of permafrost contributes to the acceleration of global warming. This melting can also cause havoc for the means of subsistence and cultural practices that indigenous groups rely on.

Variations in the Snow Cover:

The shifting patterns of seasonal snow cover can have an effect on the amount of water that is available, in particular in areas where snowmelt is a key source of freshwater. Changes in the timing of the snowmelt can result in a lack of water, which will have repercussions for agriculture, ecosystems, and communities.

Consequences for Both the Region and the World as a Whole

The changes that are taking place in the cryosphere will have repercussions that are felt far beyond the polar regions. They have an effect on the patterns of weather, the rise in sea level around the world, as well as the severity and frequency of extreme weather occurrences. For instance, the reduction in the amount of sea ice in the Arctic can change the patterns of air circulation, which in turn can have an effect on the weather systems in places located in the middle latitudes.

The melting of glaciers and ice caps poses a threat to the water supply of millions of people who rely on meltwater for drinking, farming, and hydroelectric power.

Coastal erosion and the flooding of low-lying coastal areas are both caused by increasing sea levels, which are caused by the melting of glaciers and ice caps. Additionally, the expansion of seawater as it heats also contributes to the problem.

2.2 Examination of polar ice caps, glaciers, and permafrost

The ice caps of the poles, glaciers, and permafrost are all essential parts of the Earth's cryosphere. They each play an important part in the climate and biosphere of the planet. The study of these frozen features, which cover enormous parts of the surface of the Earth, is essential to having a complete understanding of climate change, sea level rise, and ecosystem dynamics. This article examines the study of polar ice caps, glaciers, and permafrost in order to offer insight on the scientific methods and technology that are used to investigate these frozen landscapes as well as the ramifications of their changing circumstances.

The Ice Caps at the Poles

The huge areas of ice and snow that blanket the Earth's northern regions are known as polar ice caps. These enormous sheets of ice can be found in both Antarctica and the Arctic, and they play a significant part in the process of controlling the weather on Earth by acting as a reflector of sunlight and helping to keep the temperature on the globe from fluctuating too much. The study of the polar ice caps requires a multidisciplinary approach, making use of a variety of methodologies and technology to monitor the ice caps' behavior, as well as its size and mass.

Satellite-Based Remote Sensing Technology

Satellite remote sensing is one of the key ways that scientists use to investigate the polar ice caps. The Earth is orbited by satellites that are outfitted with a variety of sensors and instruments, and these satellites give essential data on the extent, thickness, and variations in ice cover over time. For instance, the CryoSat-2 mission of the European Space Agency used radar altimetry to determine the thickness of the polar ice caps. This provides specific information regarding the volume and mass loss of the ice caps.

Satellites utilize optical and infrared sensors to monitor the extent of snow cover and identify surface features such as fissures, crevasses, and variations in albedo, which shows the reflection of the ice. Measuring ice thickness is just one of the many things that these sensors can perform. For example, the Moderate Resolution Imaging Spectroradiometer (MODIS) aboard NASA's Terra and Aqua satellites provides high-resolution photography that assists in monitoring changes to ice caps. This data is used by NASA.

Cores of Ice

Drilling into the ice caps and removing core samples of ice is another important method for conducting research on the polar ice caps.

Ice cores are cylinder-shaped samples that are extracted from ice sheets and provide a record of the environment in the past where they were found. Scientists are able to reconstruct the history of the Earth's climate by analyzing the air bubbles, isotopes, and other impurities that are trapped within the ice. This includes changes in temperature, changes in atmospheric composition, and volcanic eruptions. For example, the European initiative for Ice Coring in Antarctica (EPICA) initiative collected ice cores from Antarctica, which revealed a climate history that spanned hundreds of thousands of years and covered hundreds of thousands of years.

Ice cores not only provide information on climates in the past, but they also assist in the process of calibrating and validating climate models, which are vital for projecting how climate will change in the future. In addition, ice cores help to our understanding of how polar ice caps react to changes in climate and how these responses may continue to evolve in the future.

Glaciers

Glaciers are enormous volumes of ice that flow under their own weight, sculpting landscapes and contributing to oscillations in sea level. Glaciers can be found all over the world. They can be found in many mountainous places all over the world and, similar to the polar ice caps, they are sensitive markers of how the climate is changing. In order for scientists to obtain accurate results from their research on glaciers, they employ a variety of methods, including field observations, remote sensing, and modeling.

Measurements Taken in the Field

Research on glaciers cannot be considered complete without field measurements. Researchers travel to glaciers in order to get data on the mass balance, ice movement, and melt rates of the glaciers. In order to accomplish this, equipment such as ablation stakes, snow pits, and GPS receivers must be placed across the glacier so that changes in its behavior and dimensions may be monitored. Researchers are able to evaluate the effects of climate change on the health of glaciers by comparing data taken at different times over time.

Glacier monitoring stations are an important part of the overall glacier research endeavor. These stations are outfitted with numerous monitoring devices, such as mass balance stakes, weather stations, and ice-penetrating radar, in order to maintain constant vigilance over a wide range of parameters. Scientists are able to discern patterns and oscillations in glacier behavior, such as thinning, retreating, and changes in flow speed, thanks to the long-term data collected from these stations.

Sensing from a Distance

Monitoring glaciers on a broader scale requires the use of remote sensing in order to be effective.

Images and data that disclose the extent of glaciers, their flow, and the structures on their surfaces can be captured by satellites equipped with optical and radar sensors. A comprehensive record of glacier change throughout time has been made possible thanks to the joint efforts of NASA and the United States Geological Survey in the development of the Landsat program. Researchers are able to identify changes in glacier boundaries, ice thickness, and surface temperature with the assistance of satellite photography.

Recent years have seen significant advancements in glacier research thanks to the emergence of unmanned aerial vehicles, also known as drones. High-resolution pictures, three-dimensional maps, and thermal data can be captured by drones fitted with specific sensors; these types of data are particularly helpful for monitoring the health of glaciers. The use of these equipment has made it simpler to reach inaccessible and difficult-to-navigate glacier regions.

Making models

Glacier modeling is an additional important part of glacier study. Scientists are able to better forecast how glaciers will react to future climate change by using numerical models that replicate the behavior of glaciers under a variety of different climate scenarios. In order to assess the progression of glaciers, these models take into account a variety of elements, including temperature, precipitation, topography, and glacier shape.

The Ice Sheet Model Intercomparison Project for CMIP6 (ISMIP6) is an example of one of these

modeling efforts, and its primary goal is to improve forecasts of the contributions of ice sheets to the rise in sea level. Scientists achieve a more all-encompassing understanding of glacier dynamics and their impact to changes in sea level when they combine observations with simulation.

The permafrost

The term "permafrost" refers to frozen ground that has been maintained at a temperature at or below freezing for at least two years in a row. It encompasses extensive areas of the Northern Hemisphere, such as Siberia, Alaska, and Canada, and it is an essential component in the circulation of carbon across the world. The study of permafrost requires the utilization of a variety of methods, ranging from measurements taken on the ground to observations made by satellite.

Measurements Taken From the Ground

Researchers who are interested in permafrost conduct measurements on the ground in order to determine the temperature of the permafrost, the thickness of the active layer, and the makeup of the soil. In order to monitor temperature profiles as well as changes that occur over time, boreholes and thermistor arrays are utilized.

The data gathered enables scientists to quantify the degree to which permafrost is susceptible to thawing and to evaluate the effects of this susceptibility on ecosystems and infrastructure.

A major indicator of the health of permafrost is the active layer thickness, which refers to the depth of the soil layer that experiences seasonal thawing and freezing. Researchers utilize a variety of instruments, including as frost probes and ground-penetrating radar, to quantify the thickness of the active layer and track how it varies in response to shifting climatic circumstances.

Sensing from a Distance

When conducting research on permafrost across greater areas, remote sensing is an extremely helpful technique. The Advanced Microwave Scanning Radiometer (AMSR-E) and the Soil Moisture and Ocean Salinity (SMOS) satellite are two examples of satellite-based sensors that are used to estimate the soil moisture content and the freeze/thaw state. These estimations provide insights into the conditions of permafrost found throughout huge regions.

The use of thermal images captured by satellites and drones is another helpful method for locating regions where there are considerable temperature differences. These kinds of fluctuations may be an indicator of the degradation and thawing of permafrost. Changes in surface temperature, which might be an indicator of disruptions in permafrost, can be detected with the help of infrared and thermal sensors, which are both valuable tools.

Making models

Modeling permafrost is a crucial component of permafrost research because it enables researchers to forecast how permafrost will react to changes in the climate in the future. Computer models can be used to mimic the dynamics of permafrost by taking into account variables such as air temperature, snow cover, and vegetation. These models are able to make projections on changes in the permafrost's area, depth, and carbon release under a variety of different climate scenarios.

The potential release of greenhouse gasses, particularly methane, as a result of the thawing of permafrost is one of the most important discoveries made by permafrost modeling. The release of methane into the atmosphere is a significant contributor to the acceleration of climate change. Scientists are able to better understand the function that thawing permafrost plays in the climate system and incorporate this understanding into their climate projections by utilizing models to estimate methane emissions from thawing permafrost.

The Effects That the Melting of Polar Ice Caps, Glaciers, and Permafrost Have On Humanity

Sea Level Rise: The sea level is increasing as a result of melting polar ice caps and glaciers, which poses a hazard to coastal residents and ecosystems. It is absolutely necessary for coastal planning and adaptation to have accurate monitoring and modeling of these changes.

The release of greenhouse gases from thawing permafrost can produce feedback loops in the climate, further intensifying the effects of global warming. The study of permafrost enables us to more accurately evaluate the potential dangers connected with these feedbacks.

Impacts on Ecosystems: Glaciers, ice caps, and permafrost are all extremely important components of many ecosystems and the habitats of various species of wildlife. Alterations to these icy settings have the potential to wreak havoc on food chains, biodiversity, and the services provided by ecosystems.

Glaciers act as natural water reservoirs, ensuring a steady supply of clean water for millions of people living downstream from them. For the purpose of managing water resources in many different places, having a solid understanding of glacier melt and the time of water release is essential.

Extreme Events: Rapid changes in ice caps and glaciers can lead to glacial lake outburst floods and avalanches, which provide risks to people located in close proximity to these places.

The polar ice caps are an important component in the climate-controlling process that occurs across the planet. Alterations in their reflectivity (albedo) and heat exchange with the atmosphere have the potential to have far-reaching effects on the patterns and circulation of the weather on a global scale.

Understanding the Earth's cryosphere and the tremendous impact it has on the climate and ecosystems of the globe requires a thorough investigation of polar ice caps, glaciers, and permafrost. This investigation is crucial. In order to monitor and investigate these frozen frontiers, scientists utilize a wide variety of methods and technology. Some of these include ice core analysis, satellite remote sensing, and permafrost modeling. Their research not only helps shed light on climate changes that have occurred in the past and are occurring in the present, but it also contributes to efforts to forecast what the future effects of melting ice and thawing permafrost will be. The realization of the significance of these changes, which include the rise in sea level as well as disruptions to ecosystems and feedback loops in the climate system, highlights the need of continuing research in this sector.

It is crucial that we continue to invest in the research of polar ice caps, glaciers, and permafrost in order to get a deeper understanding of the issues posed by our planet's changing state and to find effective solutions to those challenges.

2.3 Role of the cryosphere in maintaining Earth's climate balance

The cryosphere of the Earth, which consists of the polar ice caps, glaciers, and permafrost, is an essential component in the process of preserving the climate equilibrium of the planet. These frozen components of the Earth's system not only manage the transfer of heat and water, but they also have an influence on global weather patterns and the level of the sea. In this essay, we will investigate the myriad of ways in which the cryosphere contributes to the consistency of the climate on Earth, as well as the reasons why it is crucial to understand and safeguard the frozen regions on our planet.

The concepts of Reflectivity and Albedo

The high reflectivity, also known as albedo, that the cryosphere possesses is one of the fundamental mechanisms that contributes to the cryosphere's role in preserving the Earth's climate equilibrium. Both snow and ice have high albedo values, which means that they reflect a considerable percentage of the solar energy that is arriving from the outside back into space. This reflection of sunlight serves to keep the surface of the earth colder and contributes to the process of temperature regulation.

As the temperature rises, the ice and snow cover will eventually melt, revealing darker surfaces such as open water, rock, and soil. These surfaces have lower albedo values, which means they absorb more solar radiation and contribute to a further rise in temperature. Climate change is made worse by the positive feedback cycle described above, in which increased heat leads to greater ice melting and less reflectance. Therefore, the existence of a healthy cryosphere contributes to the reduction of the warming impact that is caused by greenhouse gases. This is accomplished through the reflection of solar radiation back into space.

The Redistribution of Heat Around the World

In addition to this, the cryosphere is critically important for the redistribution of heat over the world. It performs the function of a massive heat sink, soaking up extra heat during warmer seasons and releasing it when temperatures drop. This redistribution of heat contributes to the preservation of the overall temperature equilibrium that exists on Earth.

The ice that is contained within glaciers and polar ice caps contains enormous volumes of freshwater. The melting of these ice reservoirs will cause freshwater to be released into the oceans as the climate continues to rise.

This cold freshwater influx can have an effect on the patterns of ocean circulation, particularly the thermohaline circulation, which is what drives the global ocean currents. Alterations in ocean circulation can, in turn, have an effect on climate patterns by shifting the distribution of heat and affecting the functioning of weather systems.

Regulation of the Sea Level

The cryosphere's influence on sea levels is yet another important part of the climate balancing act that it plays. The ice sheets, glaciers, and ice caps that cover the poles all store enormous amounts of freshwater. If all of the ice on Earth were to melt at once, it would cause a rise in sea level of several meters around the world. On the other hand, the cryosphere serves as a brake on the rise in sea level.

During periods of time with lower temperatures, ice forms on glaciers and polar ice caps, which traps significant quantities of water in its frozen state. This ice is melting as a result of the warming environment, and it is flowing into the ocean, which contributes to the rise in sea level. There is a clear correlation between the rate of ice loss or gain in the cryosphere and the rise or fall in global sea levels. Because of this, having a knowledge of and keeping track of changes in these frozen places is vital for predicting the effects of sea level rise and finding ways to mitigate them. Sea level rise can flood coastal towns and inflict severe damage.

Influence on the Typical Weather Patterns

The distribution of temperature and moisture can be modulated by the cryosphere, which in turn influences the weather patterns and regional climates. For instance, alpine glaciers are known to have the potential to have a significant impact on the climate of the regions that they feed. When glaciers melt, they pour freshwater into rivers and streams, which ensures that there is a consistent supply of cold water. This icy glacial meltwater has the potential to bring the air temperature in the surrounding region down, so producing a variety of distinct microclimates.

In a similar vein, polar ice caps and sea ice have a considerable influence on the weather patterns that occur in regional areas. During the summer months, when sea ice is melting, it discharges cold, fresh water into the Arctic Ocean as well as the waters that surround it. This cold, less dense water has the potential to disrupt ocean circulation patterns and influence atmospheric conditions. As a result, there is a possibility that weather systems, such as the jet stream and storm tracks, would shift. These kinds of shifts in weather patterns can have far-reaching repercussions, not only in the regions close to the polar regions but also further afield.

Influence on the World's Ecosystems

The ecological systems of polar and alpine regions are both impacted by the cryosphere's presence and influence. The existence of ice and snow has an effect on the habitats, ecosystems, and patterns of migration of a wide variety of animal species. A great number of creatures have developed strategies to survive in these inhospitable settings, and the seasonal cycles of freezing and thawing are what they rely on to control their life cycles.

Polar bears and seals, for instance, rely on the sea ice as a platform for foraging and sleeping in order to survive. Their access to prey and breeding sites is determined, in part, by the time and magnitude of the sea ice melt. In a similar fashion, mountain ecosystems have adapted to the presence of glaciers and snow. Changes in the frozen landscapes that make up mountain ecosystems can upset the delicate balance of these ecosystems, which in turn can have an impact on the plant and animal species that lives there.

Mechanisms for Obtaining Feedback

1. **Ice-Albedo Feedback:** The albedo effect is a powerful feedback mechanism, as was described earlier. The melting of ice and snow exposes darker surfaces, which absorb more heat, causing the temperature to rise even further and causing additional ice to melt.

2. **The Thawing of Permafrost:** The permafrost, which is a component of the cryosphere, contains enormous quantities of organic carbon. Methane, which is a powerful greenhouse gas, is released into the atmosphere whenever there is thawing of permafrost. Emissions of methane can lead to an increase in global warming, which can then create a feedback loop.

3. Alterations to the Circulation of the Oceans Freshwater is released into the oceans as a result of glaciers and polar ice melting. This influx of freshwater has the potential to change the patterns of ocean circulation, which in turn can have an effect on climate and weather systems.

4. **The Rise in Sea Level:** The melting of ice caps and glaciers is a contributing factor to the rise in sea level. Coastal areas can become inundated as a result of rising sea levels, which can affect coastal ecosystems and cause additional land to become submerged in water.

The Cryosphere in a Climate That Is Always Changing

The cryosphere of the Earth is not immune to the effects that climate change will have on it. As a result of the increase in average temperature around the planet in recent decades, we have witnessed dramatic shifts in the polar ice caps, glaciers, and permafrost. These alterations will have a significant and far-reaching impact on the consistency of the climate system of the world.

The polar ice caps and glaciers are also melting.

Over the past few years, there has been a considerable amount of ice lost from both the polar ice caps and glaciers. The area of sea ice in the Arctic has shrunk dramatically in recent years, with the summer minimum decreasing at an alarming rate. There are concerns over the stability of the West Antarctic Ice Sheet in Antarctica, which, in the event that it were to collapse, might lead to a significant rise in the level of the ocean.

In addition, mountain glaciers all across the world are rapidly receding at an alarming rate. This melting of ice has an impact not just on the rise in sea level but also on the supply of water in areas that are dependent on glacial meltwater. Many towns get their drinking water, their agricultural supplies, and their power from rivers that are fed by glaciers.

Permafrost beginning to thaw

The thawing of permafrost is another serious worry brought about by global warming. Methane and carbon dioxide are two of the gases that are released into the atmosphere when permafrost begins to thaw. The thawing of permafrost can lead to higher emissions of the potent greenhouse gas methane, which can hasten the process of global warming. Additionally, in areas where there is a significant amount of permafrost, the thawing of permafrost can result in the destruction of infrastructure, such as roads and buildings.

The Raise in Sea Level

There is no denying the fact that glaciers and ice caps in the polar regions are melting, which is contributing to the rise in sea level. As a result of rising sea levels, coastal regions are already suffering flooding that is both more frequent and more severe. It is possible for rising sea levels to have a cataclysmic impact on low-lying coastal communities, leading to the displacement of inhabitants and causing economic and environmental harm.

Reducing the Effects That the Cryosphere's Changes Will Have

Reducing Emissions of Greenhouse Gases The increase in greenhouse gas emissions is the principal factor responsible for the warming of the planet's climate. It is essential to cut emissions by taking steps such as shifting to renewable energy sources and improving energy efficiency if we want to preserve the cryosphere and keep the climate in a stable state.

Monitoring and Research: The cryosphere requires constant monitoring and research, both of which are absolutely necessary. This involves monitoring the changes that are occurring in the polar ice caps, glaciers, and permafrost, as well as gaining an understanding of the feedback mechanisms that are at play. For the purpose of informing climate models and making forecasts, data from satellite observations, field measurements, and measurements taken from ice cores are essential.

Adaptation to Climate: Communities in susceptible locations need to adopt ways to adjust to changing conditions, including rising sea levels and permafrost thaw. This is required as part of climate adaptation. Relocating infrastructure, constructing coastal defenses that are resilient, and diversifying local economies are all potential components of these solutions.

Cooperation Across International Boundaries Climate change is a worldwide problem that calls for cooperation across international boundaries. In order to combat climate change and maintain the cryosphere, international agreements such as the Paris Agreement, which establish targets for reducing emissions of greenhouse gases, should be supported and strengthened.

Public Consciousness: It is vital to educate the public about the significance of the cryosphere and the role it plays in maintaining the climate's equilibrium. The will of political leaders can be influenced by the support of the public for climate action, which can then lead to greater measures to protect these frozen places.

The cryosphere of the Earth, which consists of the polar ice caps, glaciers, and permafrost, is an essential part of the climate system of the planet as a whole. Through mechanisms such as a high albedo, heat redistribution, and management of sea level, it plays a significant part in the process of keeping the climate on Earth in a stable state. However, despite their frozen state, these locations are extremely susceptible to the effects of climate change, as evidenced by the fast melting and thawing that has been taking place in recent years.

For effective climate action, it is essential to have a solid understanding of the importance of the cryosphere and how it will react to changing temperatures. Protecting and preserving these frozen regions, lowering emissions of greenhouse gases, and adjusting to changing conditions are all vital efforts that must be taken in order to reduce the effects of changes in the cryosphere and maintain the climate equilibrium on Earth. In addition to being an essential component of the climate system of the Earth, the cryosphere is also an essential component of our collective obligation to preserve the planet for the benefit of future generations.

3

Chapter 3

Melting Polar Ice Caps

In recent years, one of the most pressing issues of concern has been the retreat of the ice caps that cover the poles, in particular in the Arctic and Antarctic areas. As a result of human activity, the climate of the Earth is warming, which is having implications that are becoming more and more apparent as the polar ice caps continue to melt. This essay investigates the reasons for, as well as the effects of, the melting of the polar ice caps. It also investigates the scientific research that underpins this phenomenon and considers alternative options for both adaptation and mitigation.

1. **The factors that are causing the polar ice caps to melt**
   1.1. The Changing Climate
   Climate change, which is caused by an increase in the concentration of greenhouse gases in the atmosphere of the Earth, is the principal factor that is contributing to the melting of the polar ice caps. The phenomena that is known as global warming is caused by greenhouse gases such as carbon dioxide (CO_2), methane (CH_4), and others. This leads to an increase in average

temperatures around the globe. One of the most noticeable and concerning effects of this process is the melting of the ice caps that cover the poles.

1.2. Mechanisms for Receiving Feedback

Feedback mechanisms are another way in which the melting of the polar ice caps might contribute to climate change. For instance, as ice and snow melt, they expose darker surfaces like open water and bare land, both of which absorb more sunlight and heat, contributing further to the acceleration of climate change. The decrease in the amount of ice and snow cover causes the albedo, or reflectivity, of the Earth to decrease, which in turn causes an increase in the amount of heat that the Earth absorbs.

1.3. The Currents of the Ocean

Alterations in the patterns of ocean circulation can also have an effect on the rate at which polar ice caps melt. Ocean currents are responsible for the transfer of heat and have an effect on the overall temperature of the atmosphere as well as the oceans. These currents can be disrupted, which can lead to the warming of polar regions, which can speed up the melting of ice.

1.4. Patterns of Air Circulation in the Atmosphere

Patterns of atmospheric circulation, such as jet streams and other types of weather systems, are able to exert a considerable influence on the polar ice caps. Alterations in these patterns have the potential to have an effect on the average temperature, amount of precipitation, and number of storms that occur in the polar regions. The melting of ice caps and glaciers can be caused by a climatic system that has been altered.

2. The Repercussions of Polar Ice Caps Disappearing

2.1. The Increasing Height of the Sea

Rising sea levels is one of the most direct and easily observable outcomes that can result from the melting of polar ice caps. The melting of ice caps and glaciers causes an increase in the amount of freshwater that is contributed to the world's oceans, which in turn causes sea levels to rise. Coastal communities and ecosystems are under danger as a result of rising sea levels, which will cause erosion and floods, as well as the relocation of inhabitants.

2.2. Flooding along the Coast

Flooding along coastlines happens more frequently and is more severe as a result of rising sea levels and storm surges. Low-lying coastal regions, such as cities and infrastructure, are at a greater danger than higher-lying ones. Flooding has the potential to endanger lives, cause damage to property, and result in economic losses.

2.3. The Disturbance of Ecosystems

The melting of the polar ice caps will have significant repercussions for the ecosystems and the creatures that live in them. It has the potential to change the distribution of species, cause disruptions in food chains, and influence the patterns of mating and migration. As an illustration, polar bears rely on the platform provided by sea ice in order to hunt seals. As the ice melts, it becomes increasingly difficult for them to get to the prey they need.

2.4. The Acidification of the Ocean

The release of freshwater into the ocean as a result of the melting of polar ice caps can have an effect on the ocean's chemistry and contribute to the acidification of the ocean. The presence of additional freshwater causes the concentration of saltwater to decrease, which in turn makes the seawater more acidic. This has the potential to be harmful to marine life, specifically coral reefs and shellfish.

2.5. Variations in the Typical Weather Patterns

It is possible for melting polar ice caps to have an effect on weather patterns both regionally and globally. The melting of ice and snow exposes darker surfaces, which have a greater capacity to take up heat. This can cause disturbances in atmospheric circulation, jet streams, and weather systems, which may result in changes in the patterns of precipitation and the frequency of extreme weather occurrences.

2.6. The Retreat of Glaciers

As a direct result of global warming, mountain glaciers all around the world are receding at an alarming rate. Not only does the melting of glacial ice contribute to an increase in sea level, but it also has an effect on the water resources further downstream. Many towns get their drinking water, their agricultural supplies, and their power from rivers that are fed by glaciers.

2.7. The Rising Levels of Methane Emissions

When permafrost thaws, it has the potential to release methane, which is a powerful greenhouse gas. Permafrost is known to contain enormous amounts of organic carbon. The melting of the polar ice caps hastens the thawing of the permafrost, which may lead to an increase in the amount of methane emissions and further contribute to global warming.

3. Scientific Investigation and Continuous Observation

Comprehensive scientific research and monitoring activities are required in order to gain an understanding of the causes and effects of the melting of polar ice caps. In order to collect data and conduct research on the shifts that are taking place in polar regions, researchers use a wide variety of methods and technology.

3.1. Observations Made Via Satellite

Satellites that orbit the Earth and are outfitted with a variety of sensors and instruments are able to collect important data on the polar ice caps. These satellites take pictures and collect data on the amount and thickness of ice, as well as changes that have occurred over time. For instance, the CryoSat-2 mission of the European Space Agency used radar altimetry to assess the thickness of the polar ice caps and provide insights into their volume and mass loss. These measurements may be found here.

Changes in snow cover, surface temperature, and surface features such as crevasses and fissures are all tracked by satellites. This information is essential for monitoring the changes in ice caps and comprehending how they are reacting to climate change.

3.2. Ice Sheets and Cores

Ice cores extracted from glaciers and ice caps in the polar regions provide one of the most important sources of historical climate data. Ice cores can be found in glaciers. The layers of ice, air bubbles, isotopes, and contaminants that are contained within these ice cores provide a record of the climate conditions that existed in the past. Scientists are able to reconstruct the history of the Earth's climate by analyzing these cores, which includes variations in temperature, changes in the composition of the atmosphere, and volcanic eruptions.

Ice cores not only provide information on the climate of the past, but they also assist in the process of calibrating and validating climate models. These models are absolutely necessary in order to accurately predict future climate change and the effects of the polar ice caps melting.

3.3. Measurements Taken in the Field

Research on the polar ice caps cannot be considered complete without field measurements. In order to get information on ice mass balance, ice movement, and melt rates, scientists travel to polar regions. In order to monitor shifts in the proportions and behavior of the ice cap, various pieces of apparatus, such as ablation stakes, snow pits, and GPS receivers, are utilized. Researchers are able to evaluate the effects of climate change on the state of the ice caps by comparing data taken at different times.

Monitoring stations located on polar ice caps are also extremely important for the collecting of continuous data. Instruments such as mass balance stakes, meteorological stations, and ice-penetrating radar are installed in these sites. Using the long-term data collected at these stations, scientists are able to identify patterns and shifts in the behavior of ice caps, such as thinning, retreating, and variations in flow speed.

3.4. The Use of Models

It is possible to simulate the behavior of polar ice caps and glaciers under a number of different climate scenarios by using numerical models. In order to estimate the progression of ice caps, these models take into account a variety of elements, including temperature, precipitation, topography, and glacier geometry. For instance, the Ice Sheet Model Intercomparison Project for CMIP6 (ISMIP6) is centered on the improvement of forecasts of the contributions of ice sheets to the rise in sea level.

The potential release of greenhouse gases, particularly methane, from thawing permafrost is another topic that is investigated through modeling efforts. The release of methane into the atmosphere is a significant contributor to the acceleration of climate change. Methane emissions from thawing permafrost can be estimated with the use of models, which also contribute to our comprehension of the function that this process plays in the climate system.

4. Strategies for Mitigation as well as Adaptation

4.1. Cutting Down on Emissions of Greenhouse Gases

The reduction of emissions of greenhouse gases is the most effective solution to the problem of melting polar ice caps. This requires making the switch to renewable energy sources, increasing energy efficiency, and establishing rules to limit emissions from many sectors, including transportation, industry, and agriculture, among others.

4.2. Agreements Reached on a Global Scale

Cooperation on a global scale is very necessary in order to combat climate change and mitigate the negative effects of the retreating polar ice caps. Targets for the reduction of greenhouse gas emissions and the promotion of global efforts to prevent global warming can be established through international agreements such as the Paris Agreement. In order to be successful in the fight against climate change, such accords need to be maintained and strengthened.

4.3. Adaptation to the Climate

Adaptation plans are something that communities in susceptible locations need to work on developing in order to survive changes in their environment, such as rising sea levels and increasing floods along the coast. Building coastal fortifications that are resistant to erosion, shifting critical infrastructure, and diversifying local economies are all potential components of these plans.

4.4. Observations and Investigations

Continued monitoring and study are essential for creating effective methods for mitigating the effects of melting polar ice caps as well as for understanding the repercussions of melting polar ice caps. This involves monitoring the changes that are occurring in the polar ice caps, glaciers, and permafrost, as well as gaining an understanding of the feedback mechanisms that are at play.

4.5. Awareness of the Public

It is absolutely necessary to work towards increasing public understanding of the significance of addressing melting polar ice caps and climate change. The will of political leaders can be influenced by the support of the public for climate action, which can then lead to greater measures to protect these frozen places.

One of the most significant effects of global warming and climate change is the disintegration of the polar ice caps. It has been determined that a changing climate is the key factor in causing this occurrence. These causes have been thoroughly investigated.

The effects will be felt far and wide, having an impact not only on sea levels but also on coastal communities, ecosystems, and weather patterns. Data and insights about the changes taking place in polar regions can be gleaned from valuable research and monitoring initiatives, such as satellite observations, ice core analysis, field measurements, and modeling.

It will take concerted efforts to cut greenhouse gas emissions, establish international agreements, adapt to changing conditions, and promote public awareness in order to mitigate the effects of melting polar ice caps. The preservation of the cryosphere is necessary not only for the health of the planet as a whole, but also for the continuation of eco-systems and communities that are dependent on it. In order to protect the world and the generations that will follow after us, it is imperative that immediate action be taken to address the severe problem posed by the melting of the polar ice caps.

3.1 In-depth exploration of the Arctic and Antarctic regions

The Arctic and Antarctic areas, which are geographically located on opposite sides of the planet, are two of the world's most inaccessible, untouched, and difficult regions to visit. These immense frozen land-scapes attract the imaginations of explorers, scientists, and other types of adventurers and travelers. In this article, we are going to begin on an in-depth examination of the Arctic and Antarctic areas, digging into their distinctive qualities, the problems and opportunities that they bring, and the significance of understanding these habitats in the context of climate change and global ecosystems. Specifically, we will be focusing on the Arctic and Antarctic regions' unique traits, as well as the challenges and opportunities that they present.

The Arctic Part of the World

The landmasses that lie within and immediately to the north of the Arctic Circle are a part of the Arctic region, which also includes the ocean that is to the north of it and the seas that surround it. It is defined by its frigid waters, its harsh climate, and the immense areas of sea ice and permafrost that cover the land.

The weather and the map both.

The climate of the Arctic is severe, with bitterly cold temperatures and lengthy, gloomy winters, followed by fleeting, intense summers. The region is affected by a phenomena known as polar amplification, which indicates that it warms at a rate that is roughly twice as fast as the average temperature for the entire planet. This increased warming has significant repercussions for the climatic system that exists in the Arctic.

The majority of the Arctic Ocean is covered with sea ice, which grows and shrinks with the passage of time and the different seasons.

The Arctic Ocean is almost entirely frozen throughout the winter months, and sea ice can be found covering great swaths of the ocean's surface. During the summer, the ice cover melts, exposing the open water and resulting in a habitat that is always evolving.

Animal and plant life

The Arctic is home to a surprising variety of living forms despite the harsh conditions that are typically found there. The polar bear, the Arctic fox, and the reindeer are just few of the iconic species that have adapted to the frigid temperature and lengthy winters. There is a rich diversity of marine life in the waters close by, including whales, seals, and a wide variety of fish species. In addition, there is a great diversity of bird species in the Arctic, and the region is an essential breeding place for a large number of migratory birds.

The tundra biome encompasses a significant portion of the Arctic region and is distinguished by frigid climates, barren landscapes, and vegetation that grows at a low elevation. In this hostile climate, the most common types of plant life are lichens, grasses that can withstand the conditions, and mosses.

Investigation as well as Exploration

Early explorers in the Arctic searched for the fabled Northwest Passage, which is a sea path that passes over the Arctic Ocean and connects the Atlantic and Pacific oceans. This region has a long and illustrious history of exploration. Roald Amundsen, Robert Peary, and Richard E. Byrd are just a few of the well-known explorers who were the first to travel to this region and open it up to scientific research and human activities.

Studies on climate change, ice dynamics, marine habitats, and the effects of global warming are just few of the topics that are currently being investigated through scientific research in the Arctic today. To examine the region's shifting conditions and gain a better understanding of the crucial role it plays in the climate system of the Earth, scientists make use of a wide range of equipment, such as icebreakers, research stations, and technologies that utilize remote sensing.

The Arctic is responsible for a sizeable portion of the planet's overall temperature regulation. Its high albedo, also known as its reflectivity, means that it reflects a significant percentage of the solar energy that is arriving at Earth back into space, which in turn helps to chill the planet. However, when the region heats and the sea ice melts, the albedo of the region will decrease. This will result in higher heat absorption and rapid warming. This is a feedback mechanism that will have far-reaching effects for the climate of the entire planet.

Traditional Societies and Groups

Indigenous peoples from all over the world have made their homes in the Arctic, including the Inuit, Saami, Nenets, and a great many others. These groups have a long history of habitation in the area, during which time they have maintained their way of life through the use of time-honored methods of hunting, fishing, and herding. These communities are being hit hard by the effects of climate change because it is causing disruptions in the ecosystems that they rely on, which in turn is contributing to difficulties in areas such as food security, cultural preservation, and adaptability.

The Antarctic Part of the World

In contrast to the Arctic, the Antarctic region consists of a massive continent that is surrounded on all sides by the Southern Ocean. It is the place on Earth that is the windiest, driest, and coldest, and it is also home to the South Pole, a geographical landmark that has captivated explorers and scientists for more than a century.

The weather and the map both

The Antarctic continent is characterized by extremely low temperatures and a climate similar to that of a polar desert. Temperatures in the interior can fall to shockingly low levels, falling as low as -128.6 degrees Fahrenheit (-89.2 degrees Celsius) during the long, dark winter. Temperatures tend to be more bearable in coastlines because the ocean that surrounds these areas helps to moderate the climate.

The East Antarctic Ice Sheet is the single most important feature of Antarctica's geography. It is home to the vast majority of the world's freshwater ice. The region of West Antarctica is distinguished by a number of glaciers and ice streams that move in a westerly direction toward the ocean. The continent is encircled on all sides by the Southern Ocean, which is an important factor in the regulation of climate patterns across the globe.

Animal and plant life

Because of Antarctica's severe weather, the continent's terrestrial ecosystems remain relatively undeveloped. There are just a select few plant species that are able to endure the extreme cold and dry circumstances, and some of those are lichens, algae, and mosses. On the other hand, despite the fact that the Southern Ocean is covered in ice, there is an astounding variety of marine life that thrives within its freezing waters.

The Southern Ocean is home to a diverse collection of marine life, such as krill, penguins, seals, and many other kinds of fish. As a result of the fact that the surrounding waters constitute an essential feeding site for whales, the region is of the utmost importance for the conservation of marine life.

Investigation as well as Exploration

Legendary characters such as Ernest Shackleton, Robert Falcon Scott, and Roald Amundsen played significant roles in the history of Antarctic exploration. In their pursuit to reach the South Pole and get a better understanding of Antarctica, these intrepid explorers embarked on dangerous trips to the continent, which frequently took place in dangerous conditions.

The scientific investigation that takes place in Antarctica is comprehensive and covers a wide range of fields, such as astronomy, glaciology, biology, and climate science. The Amundsen–Scott South Pole Station and the McMurdo Station are two examples of research stations that act as bases for many kinds of scientific inquiries. As a result of its arid and consistent climate, Antarctica serves as an important center for research into space and the study of cosmic rays.

The proof that Antarctica is experiencing a rapid climatic change is one of the most important discoveries made by scientists there. Ice cores extracted from the Antarctic ice sheet include a historical record of past climate conditions. This record reveals patterns of warming and cooling that have occurred over the course of millennia. Ice cores were dug from the Antarctic ice sheet. Researchers have made use of this knowledge to develop their understanding of the dynamics of the climate and to develop their forecasts for the future.

The Protection of the Environment

The Antarctic Treaty, an international agreement that defines Antarctica as a place of peace and scientific cooperation, serves as the legal framework for the administration of the Antarctic territory. The pact emphasizes the significance of conducting scientific research and preventing damage to the environment by prohibiting activities such as mining minerals, conducting nuclear tests, and engaging in military operations on the continent.

The Protocol on Environmental Protection to the Antarctic Treaty, more generally referred to as the Madrid Protocol, establishes Antarctica as a natural reserve dedicated to peace and science. This protocol is commonly known as the Madrid Protocol. It set stringent standards for environmental protection, ensuring that human activities in the region have an impact that is as little as possible on the ecology.

Consequences of the Disappearance of Polar Ice Caps

The melting of the polar ice caps has significant repercussions for the temperature patterns around the world, the rise in sea level, and the functioning of ecosystems. To effectively address the issues posed by climate change, it is vital to have a solid understanding of the implications of the melting of ice caps.

Increasingly High Sea Levels

The rise in sea levels is the most direct and obvious result that can be attributed to the melting of polar ice caps. The melting of ice caps and glaciers causes an increase in the amount of freshwater that is added to the oceans, which in turn causes the sea level to rise. This poses a huge risk to the people and ecosystems that are found along the coast, and it could lead to the displacement of populations as well as erosion and flooding.

Flooding along the Coast

Flooding along coastlines happens more frequently and is more severe as a result of rising sea levels, which also make storm surges stronger. Particularly susceptible to damage are low-lying coastal locations, which might include cities and infrastructure. Flooding along the coast can endanger people's lives, cause damage to property, and result in economic losses.

Ecosystems in Danger of Being Disturbed

The melting of the polar ice caps will have significant repercussions for the ecosystems and the creatures that live in them. It has the potential to change the distribution of species, cause disruptions in food chains, and influence the patterns of mating and migration. Changes in the amount of sea ice and the temperature of the ocean have a direct impact on iconic species such as polar bears, seals, penguins, and whales.

Acidification of the Ocean

The release of freshwater into the ocean as a result of the melting of polar ice caps has the potential to modify ocean chemistry and contribute to the acidification of the ocean. The presence of additional freshwater causes the concentration of saltwater to decrease, which in turn makes the seawater more acidic. The acidity of the ocean has the potential to destroy marine life, such as coral reefs and shellfish, as well as to destabilize the entire marine food web.

Alterations to the Typical Weather Patterns

The polar ice caps have the potential to affect weather patterns on both a regional and global scale as they melt. The melting of ice and snow exposes darker surfaces, which have a greater capacity to take up heat. This can cause disturbances in atmospheric circulation, jet streams, and weather systems, which may result in changes in the patterns of precipitation and the frequency of extreme weather occurrences.

Influences on the Climate of the World

The polar regions are responsible for a significant portion of the climate regulation that occurs across the world. The high albedo, or reflectivity, of ice and snow helps to chill the Earth by reflecting a significant portion of the sun's incoming radiation back into space.

This reduces the amount of heat that is absorbed by the planet. A feedback process that has far-reaching implications for the global climate is one in which the polar regions warm up and the ice that covers them melts, causing their albedo to drop. This, in turn, leads to higher heat absorption and rapid warming.

Authorization of the Release of Greenhouse Gases

The release of greenhouse gasses from the thawing of permafrost can be caused by the melting of the polar ice caps. When permafrost thaws, it produces methane, which is a powerful greenhouse gas; this is because the permafrost contains enormous amounts of organic carbon. Emissions of methane can contribute to additional warming of the planet, which can then create a feedback loop that increases climate change.

Influence on Various Native American Communities

Indigenous communities in both the Arctic and the Antarctic will be directly impacted by the impacts of the melting of the polar ice caps. These people are dependent on centuries-old methods of hunting, fishing, and herding, all of which are being made more difficult by climate change. As ecosystems undergo shifts, the difficulties associated with maintaining cultural traditions, ensuring food security, and adjusting to change become more significant.

Both Mitigation and Adaptation are Needed

In order to address the issues that may arise as a result of the polar ice caps melting, it will be necessary to implement both mitigation initiatives to cut greenhouse gas emissions and adaptation methods to deal with the changes that are currently occurring. The following is a list of significant techniques that can be used to minimize and adapt to the impacts of melting ice caps:

Bringing Down Emissions of Greenhouse Gases

The most efficient approach to dealing with the problem of melting polar ice caps is to lower emissions of greenhouse gases. Important actions to take in the fight against climate change include making the switch to renewable energy sources, increasing energy efficiency, and putting into place regulations to put a cap on emissions from a variety of industries.

Agreements reached on a global scale

For the purpose of combating climate change and mitigating the effects of the ice caps in the polar regions melting, international collaboration is very necessary. Conventions such as the Paris Agreement establish goals for the reduction of emissions of greenhouse gases and encourage collective efforts to curb the effects of global warming. In order to be successful in the fight against climate change, such accords need to be maintained and strengthened.

Adjustment to the Climate

Adaptation plans are something that communities in susceptible locations need to work on developing in order to survive changes in their environment, such as rising sea levels and increasing floods along the coast. Building coastal fortifications that are resistant to erosion, shifting critical infrastructure, and diversifying local economies are all potential components of these plans.

Observation and Investigation

Continued monitoring and study are essential for creating effective methods for mitigating the effects of melting polar ice caps as well as for understanding the repercussions of melting polar ice caps. This involves monitoring the changes that are occurring in the polar ice caps, glaciers, and permafrost, as well as gaining an understanding of the feedback mechanisms that are at play.

Awareness among the public

It is absolutely necessary to work towards increasing public understanding of the significance of addressing melting polar ice caps and climate change. The will of political leaders can be influenced by the support of the public for climate action, which can then lead to greater measures to protect these frozen places.

The polar regions of the Earth, including the Arctic and Antarctic, are some of the most fascinating and difficult to explore. Due to their one-of-a-kind qualities, varied ecologies, and harsh settings, they are important topics for the study and research of in the scientific community. The melting of the polar ice caps will have far-reaching implications, which will include an effect on sea levels as well as coastal communities, ecosystems, and weather patterns.

It is very necessary, in order to handle the difficulties that are given by a changing climate, to acquire an understanding of the relevance of the polar regions in relation to climate change and world ecosystems. Strategies for both mitigating the effects of melting polar ice caps and adapting to those consequences will be required in order to conserve these unspoiled and remote ecosystems for future generations. As we continue our exploration of the Arctic and Antarctic, it is imperative that we acknowledge the significance of these places and the responsibility that we all have to preserve our planet.

3.2 Causes and consequences of polar ice melt

One of the most visually apparent and alarming effects of global climate change is the melting of polar ice caps and glaciers. This is a result of global warming. This phenomena is caused by a complicated interaction of causes, and it has far-reaching repercussions for the temperature of the Earth, ecosystems on the planet, and human societies. In this essay, we will investigate the reasons for, as well as the effects of, the melting of polar ice, with particular attention paid to the Arctic and Antarctic areas.

Reasons for the Melting of Polar Ice
Changes in Climate

The principal factor that contributes to the melting of polar ice is climate change on a worldwide scale. This change is largely related to human activities, such as the combustion of fossil fuels, the clearing of forests, and industrial operations. The greenhouse effect is caused when greenhouse gases like carbon dioxide (CO_2) and methane (CH_4) are released into the atmosphere. This impact causes heat to be trapped, which ultimately leads to an increase in the average temperature of the earth.

The polar areas, and the Arctic in particular, are extremely vulnerable to the effects of changes in the global climate. Because of this sensitivity, which is referred to as polar amplification, the polar areas are warming at a rate that is nearly twice as fast as the average rate for the rest of the planet. This increased warming has a direct effect on the stability of the ice found in the polar regions.

Feedback on the Albedo

The degree to which a surface reflects light is referred to as its albedo. Snow and ice have a high albedo, which means that they reflect a sizeable percentage of the solar energy that is arriving from outside back into space. Because of its reflecting property, the surface of the Earth stays a little bit cooler. The polar ice caps and glaciers, however, start to melt as the climate warms, which exposes darker surfaces such as open water and naked land. These darker surfaces have lower albedo values, which means they absorb more solar radiation than lighter surfaces, which leads to increased warming. Climate change is made worse by the positive feedback loop described above, in which increased warming leads to further ice melting, which in turn leads to further increased warming.

Currents of the Ocean

The patterns of ocean circulation are a substantial contributor to the melting of polar ice. Ocean currents that are already heated have the ability to transport heat to the polar regions, which can contribute to the warming of ocean waters and the melting of sea ice.

Alterations in the circulation patterns of the ocean can cause an influx of warmer water, which can speed up the process by which ice melts.

Circulation of the Atmosphere

Melting of polar ice can also be affected by the circulation patterns in the atmosphere, such as the jet stream and the various weather systems. Alterations in these patterns can cause variations in temperature, precipitation, and the frequency of storms in polar regions, which can ultimately lead to changes in ice caps and glaciers.

Activities Involving Humans

It is possible for human activities in polar regions, such as research, tourism, and industrial operations, to have a localized impact on the ice in those places. For example, pollution and the release of black carbon, often known as soot, can cause the surface of snow and ice to become darker, which lowers their albedo and speeds up the melting process. The construction of new roads and buildings, among other types of infrastructure, can also cause local circumstances to shift and contribute to the melting of ice.

The Repercussions of Melting Polar Ice

Increasingly High Sea Levels

The elevation of sea levels is one of the most major and direct results that may be expected from the melting of polar ice. The melting of ice sheets, glaciers, and sea ice all add freshwater to the world's oceans, which in turn causes sea levels to rise. Coastal populations, low-lying islands, and sensitive coastal ecosystems face a significant threat as sea levels continue to rise due to climate change.

Flooding along the Coast

The rise in sea levels, in conjunction with an increase in the frequency and severity of storm surges, has the effect of making coastal flooding more severe. Cities and coastal communities that are located on lower ground are especially susceptible to flooding. Flooding along the coast can endanger people's lives, cause damage to property, and result in economic losses.

The Disruption of Ecosystems

The melting of polar ice has significant repercussions for the ecosystems and the creatures that live in polar regions. It has the potential to change the distribution of species, cause disruptions in food chains, and influence the patterns of mating and migration. Changes in the amount of sea ice and the temperature of the ocean have a direct impact on iconic species such as polar bears, seals, penguins, and whales.

Acidification of the Ocean

The introduction of freshwater into the oceans as a result of the melting of polar ice can have an effect on the ocean's chemistry and contribute to a process known as ocean acidification. The presence of additional freshwater causes the concentration of saltwater to decrease, which in turn makes the seawater more acidic. The acidity of the ocean has the potential to destroy marine life, such as coral reefs and shellfish, as well as to destabilize the entire marine food web.

Alterations to the Typical Weather Patterns

The polar ice caps have the potential to affect weather patterns on both a local and a global scale if they continue to melt. The melting of ice and snow exposes darker surfaces, which have a greater capacity to take up heat. This can cause disturbances in atmospheric circulation, jet streams, and weather systems, which may result in changes in the patterns of precipitation and the frequency of extreme weather occurrences.

Authorization of the Release of Greenhouse Gases

Large quantities of organic carbon can be found within the polar ice caps and the permafrost found in the Arctic and Antarctic regions. Methane and carbon dioxide are two powerful greenhouse gases. As these frozen regions thaw as a result of rising temperatures, they release methane and carbon dioxide into the atmosphere. This has the potential to initiate feedback loops, which would further exacerbate global warming.

Migration in Response to Rising Sea Level

People living in low-lying coastal areas are being forced to relocate as a result of rising sea levels and coastal flooding, giving rise to a phenomenon known as "climate refugees." Because of the advancement of the sea, these individuals have little choice but to find new homes.

Migration brought on by climate change can put a burden on resources, endanger people's livelihoods, and create social and political problems.

Effects on the Economy

There will be huge repercussions for the economy as a result of the melting of polar ice. Economic losses can be attributed to a number of factors, including damage to coastal infrastructure, decreased land value in vulnerable locations, and increased insurance premiums brought on by flooding. Fishing and tourism are two examples of industries that are vulnerable since they are dependent on the consistent conditions of polar habitats.

Strategies for Mitigation as well as Adaptation

In order to address the impacts of the melting of polar ice, a combination of policies, including mitigation initiatives to reduce emissions of greenhouse gases and adaptation strategies to deal with changes that are already occurring, is required:

Bringing Down Emissions of Greenhouse Gases

Reducing emissions of greenhouse gases is the first step in mitigating the effects of the melting of polar ice. This requires making the switch to renewable energy sources, increasing energy efficiency, and establishing rules to limit emissions from many sectors, including transportation, industry, and agriculture, among others.

Agreements reached on a global scale

For the purpose of combating climate change and mitigating the effects of the melting of polar ice, international collaboration is very necessary. In order to address climate change in an efficient manner, international agreements such as the Paris Agreement, which establish targets for lowering emissions of greenhouse gases, should be supported and strengthened.

Adjustment to the Climate

Adaptation plans are something that communities in susceptible locations need to work on developing in order to survive changes in their environment, such as rising sea levels and increasing floods along the coast. Building coastal fortifications that are resistant to erosion, shifting critical infrastructure, and diversifying local economies are all potential components of these plans.

Observation and Investigation

Continued monitoring and study is essential for creating effective measures for mitigating the effects of melting polar ice as well as for understanding the repercussions of melting polar ice. This involves monitoring the changes that are occurring in the polar ice caps, glaciers, and permafrost, as well as gaining an understanding of the feedback mechanisms that are at play.

Awareness among the public

It is absolutely necessary to educate the general population about the significance of taking action to combat melting polar ice and climate change. The will of political leaders can be influenced by the support of the public for climate action, which can then lead to greater measures to protect these frozen places.

The gradual disappearance of glaciers and ice caps at the poles is a direct result of the warming of the planet's climate, which is caused by a complicated web of interrelated variables.

The effects will be felt far and wide, having an impact not only on sea levels but also on coastal communities, ecosystems, and weather patterns. In order to address the challenges that are posed by the melting of polar ice, it will be necessary to combine efforts to mitigate climate change by lowering emissions of greenhouse gases with initiatives to adapt to the changes that are already in motion.

As we observe the effects of the melting of polar ice, it is essential that we acknowledge the significance of these places in relation to climate change and the overall state of the environment around the world. The preservation of polar habitats is not only necessary for the health of the planet as a whole, but also for the continuation of ecosystems and the maintenance of human cultures' standard of living. In the midst of the problems posed by a changing climate, tackling the melting of polar ice must become a global priority in order to protect our planet and the generations that will come after us.

3.3 Impact on ecosystems and indigenous communities

The effects of the melting of polar ice are not limited to the changes in physical conditions that will occur in the Arctic and Antarctic areas. They have a significant and negative impact on the ecosystems as well as the native communities that have relied on these icy environments for hundreds of years. In this section, we are going to investigate the implications for these important parts of our environment.

Influence on the World's Ecosystems:

The melting of polar ice has a negative impact on the ecosystems of the Arctic and Antarctic regions, which in turn has an effect on the complex food chains that are necessary for the survival of animals. There are a wide variety of organisms that are dependent on the presence of sea ice for their capacity to feed and reproduce. These organisms range in size from zooplankton to polar bears. The distribution of these species shifts as a result of the melting ice, which can also cause imbalances in the food web and have an effect on the entire ecosystem.

Loss of Habitat The rapid melting of ice results in the loss of essential habitat for a wide variety of species, in particular those that have adapted to living in habitats covered in ice. Seals, for instance, give birth and rest on platforms made of sea ice, and polar bears kill seals when they are on the ice. These species are being pushed to adapt as the ice melts, which may result in decreased reproductive success and increased competition for resources that are becoming increasingly scarce.

Ocean Acidification The release of freshwater into the ocean as a result of the melting of polar ice caps can cause changes in the ocean's chemistry, which in turn contributes to ocean acidification.

This has the potential to have a negative impact on marine life, particularly coral reefs and shellfish, and to disrupt the entire marine food web. The effects of ocean acidification are being felt not only in the ecosystems of the polar regions but also in other parts of the world.

Impact on the Communities of Indigenous Peoples:

Threats to Traditional Lifestyles: Indigenous groups in the Arctic, such as the Inuit, the Saami, and the Nenets, have flourished for millennia by relying on traditional hunting, fishing, and herding traditions. However, these practices are coming under attack. Food security and the maintenance of cultural traditions are two important reasons why these societies are dependent on the integrity of arctic ecosystems. As the ice melts, it causes disruptions in the availability and distribution of essential species, which places a strain on the typical lifestyles of those species.

Displacement & Relocation As a result of rising sea levels and coastal flooding caused by the melting of glaciers, indigenous people are being forced to confront the necessity of relocating. These movements, which are caused by climate change, can put a pressure on resources, cause disruptions to livelihoods, and lead to social and political difficulties. In several instances, entire communities are contemplating the challenging choice of relocating to higher ground, away from the grounds on which their ancestors once lived.

Loss of Cultural Heritage Indigenous tribes have profound cultural connections to the lands they inhabit and the environment that surrounds them, yet these ties are in danger of being severed. The protection of cultural assets and indigenous knowledge is being put in jeopardy as a result of melting ice and shifting ecosystems. These communities are experiencing deep psychological and social effects as a direct result of the loss of their cultural identity and connection to the land.

Problems Associated with Adaptation One of the most major problems that indigenous groups face is adapting to situations that are always shifting. They have to come up with plans to deal with changing ecosystems, find solutions to problems with food security, and build up their resistance to the effects of climate change. Because of this, the implementation of successful adaptation techniques requires help and resources.

Chapter 4

Shrinking Glaciers

Glaciers are essential components of the Earth's cryosphere because they are enormous volumes of ice that develop over the course of many years as a result of the accumulation and compacting of snow. Nevertheless, during the past few decades, these magnificent structures have been quickly diminishing, which has contributed to dramatic changes in the patterns of the world climate. This essay investigates the reasons for, as well as the effects of, the receding of glaciers. It also investigates the scientific research that underpins this phenomenon and considers the implications that this phenomenon has for the climate of the Earth and for human societies.

1. **The Workings of Glaciers and Their Dynamics**

   Glaciers are formed when snow gradually turns into ice over an extended period of time as it accumulates and becomes more compact. As more snow falls, the weight and pressure of the snow press down on the layers beneath it, turning the snow into ice that is compact and granular in appearance. Glaciers move because of the force of gravity, and this movement results in a process known as glacial flow or deformation. Glaciers flow downward slowly. The processes that occur within glaciers require a careful balancing act between ice accumulation and ablation, which refers to the loss of ice due to melting and sublimation. Glaciers will advance and develop in size whenever the rate of accumulation is greater than the rate of ablation. On the other hand, glaciers recede and diminish when the rate of ablation is greater than the rate of accretion. This equilibrium is extremely vulnerable to shifts in temperature and precipitation, as well as to changes in the state of other environmental elements.

2. Reasons for the Retreat of Glaciers

2.1. The Warming of the Planet

The principal factor contributing to glacial retreat is global warming, which is caused by a rise in the amount of greenhouse gases present in the atmosphere of the Earth. Carbon dioxide (CO_2), methane (CH_4), and other greenhouse gases are released into the atmosphere whenever fossil fuels are burned, forests are cut down, or industrial processes are carried out.

These activities all contribute to an increase in the average temperature of the earth. This warming has an effect on glaciers all throughout the planet, forcing those glaciers to melt at a faster rate.

2.2. Feedback on the Albedo

When glaciers melt, they expose darker surfaces such as rock and soil, both of which have a lower albedo (reflectivity) than ice and snow do. This results in less solar radiation being reflected back into space. These darker surfaces absorb more solar radiation, which ultimately leads to increased heating and accelerates the melting of glaciers. Because of this positive feedback loop, the effects of global warming on glaciers are becoming more severe.

2.3. The Circulation of the Atmosphere

Alterations in the patterns of atmospheric circulation, such as alterations in wind currents and weather systems, are another factor that can impact the retreat of glaciers. The mass balance of glaciers is affected by changes in precipitation patterns, temperature gradients, and air pressure, all of which contribute to the glaciers' retreat.

2.4. The Circulation of the Ocean

Temperature increases in the ocean can have an effect on the movement of glaciers, particularly those glaciers that flow into the ocean. Warmer waters can cause the undersides of these glaciers to melt, which accelerates the glaciers' retreat and causes them to become thinner. The rate of ice loss from these marine-terminating glaciers is further affected by shifts in ocean currents as well as temperature changes in the water.

3. The Consequences of Retreating Glaciers

3.1. The Raise in Sea Level

The contribution of glacial retreat to the overall rise in sea level is one of the most significant implications of glacial retreat. The melting of glaciers causes an increase in the amount of water that is released into the oceans, which in turn causes the sea level to rise. This poses a substantial risk to low-lying coastal areas, islands, and towns that are already vulnerable, and it could result in the erosion of coastal areas, flooding, and the relocation of inhabitants.

3.2. The Effects on the Available Water Resources

Glaciers act as natural reservoirs because they store water in the form of ice and snow. Glaciers can be found all over the world. When they get smaller, it reduces the amount of freshwater that is available for use in agriculture, downstream settlements, and power generation from hydroelectric dams.

A reliable source of water is provided by glacier-fed rivers in many areas, which is especially important during the dry seasons. The melting of glaciers can cause a shortage of water and an increase in the amount of competition for the available water resources.

3.3. The Disturbance of Ecosystems

Glaciers are home to a wide range of plants and wildlife that have developed unique adaptations to survive in the extreme cold and abrasive conditions of their surroundings. The melting of glaciers is causing changes in the habitats of these species, which in turn is causing shifts in the biodiversity and distribution of the species. This disturbance has the potential to have far-reaching repercussions for the entire ecosystem, including affects on the cycling of nutrients, food chains, and ecological processes.

3.4. Variations in the Typical Weather Patterns

The melting of glaciers has the potential to alter weather patterns on both a regional and global scale. Glaciers are extremely important in terms of their ability to influence local climate conditions by controlling temperature and precipitation. If they were to vanish, it might cause shifts in temperature gradients, alterations in the circulation of the atmosphere, and an increase in the frequency of extreme weather events like heatwaves, storms, and droughts.

3.5. Potential Geological Dangers

The melting of glaciers can set off a number of different types of geological disasters, such as glacial lake outburst floods (GLOFs) and landslides. It is possible for glacial lakes to become unstable and prone to unexpected draining, which can lead to catastrophic flooding further downstream. Glacial lakes are formed when glaciers melt and release their water. In addition, the melting of glaciers can cause mountain slopes to become unstable, which in turn raises the probability of landslides and rockfalls occurring.

4. Scientific Investigation and Continuous Observation

Comprehensive scientific study and monitoring activities are required in order to gain an understanding of the mechanisms and repercussions associated with glacial retreat. In order to collect data and conduct an analysis of the shifts that are taking place in glacier systems, researchers use a variety of methods and technology.

4.1. Measurements Made in the Field

Field measurements are an extremely important part of glaciology study since they allow for direct observations of glacial dynamics, mass balance, and ice flow. Changes in glacier volume, surface elevation, and flow velocities can be monitored by scientists with the help of various pieces of equipment like stakes, boreholes, and GPS receivers.

Data collected in the field is necessary for confirming and calibrating the results obtained from remote sensing and modeling.

4.2. Sensing from a Distance

Glaciers may be monitored on both a regional and global scale using a technique called remote sensing that is based on satellites. Images and data on the glacier's surface features and the changes that have occurred over time are gathered by satellite sensors. Insights on glacier activity, mass balance, and trends in glacial retreat can be gained through the use of techniques such as satellite altimetry, synthetic aperture radar (SAR), and optical photography.

4.3. The Analysis of Ice Cores

Ice cores extracted from glaciers give a wealth of information on previous climate conditions, including variations in temperature, air composition, and environmental changes that have occurred over the course of hundreds or even millennia. Scientists are able to recreate historical climate patterns and get an understanding of the causes of glacial changes through the examination of the physical and chemical features of ice cores. This is done within the context of natural climate variability and the effect of anthropogenic factors.

4.4. Monitoring of Glaciers' Contributions to the Mass Balance

It is imperative to perform ongoing monitoring of glacial mass balance in order to properly evaluate the state of health and stability of glacier systems. The reaction of glaciers to a changing environment can be better understood through the use of mass balance measurements. These measurements estimate the net gain or loss of ice during a certain time period. The data collected from glacier mass balances allow researchers to evaluate the relative contributions of various elements to glacial retreat. These parameters include temperature, precipitation, and albedo.

4.5. The Modeling

Simulation of glacial processes and forecasting of future changes using different climate scenarios are both possible through the use of numerical modeling. Glacier models are used to simulate the behavior of glaciers over an extended period of time by incorporating data on temperature, precipitation, topography, and ice dynamics. These models assist scientists in comprehending the intricate relationships that exist between climatic variables and glacial reactions. As a result, they are able to make more accurate estimates regarding the retreat of glaciers.

5. Implications for the Climate of the World

The melting of glaciers has important repercussions for the climate system of the Earth.

The melting of glaciers, which sends freshwater into the oceans, contributes to rising sea levels, which in turn has an effect on the weather patterns that occur all across the world. The impacts of glacial retreat are felt not just in the polar areas, where the glaciers themselves are located, but also in most of the rest of the world.

5.1. The Raise in Sea Level

Along with the thermal expansion of seawater caused by warming, glacial meltwater is a substantial contributor to the rise in sea level. If glaciers continue to recede at their current rate, more water will be released into the oceans, which would result in an increase in sea level. This can lead to the erosion of coastlines, the flooding of low-lying areas, and the complete submersion of some areas. In addition, a rise in sea level can alter coastal ecosystems and the movement of ocean currents.

5.2. The Circulation of the Ocean

It is possible for glacial meltwater to affect the circulation patterns of the ocean due to the water's lower temperature and lower salinity. This can have an effect on the flow of warm and cold water masses, as well as repercussions for the climate systems of both regional and global scales. Alterations in the circulation patterns of the ocean can cause changes in the distribution of heat, which in turn can influence weather patterns and contribute to the intensification of extreme events.

5.3. Climate Feedbacks

The melting of glaciers is one factor that feeds into a number of other climatic feedback mechanisms. The melting of ice leads to the exposure of darker areas, which lowers the albedo of the Earth's surface. This results in an increase in heat absorption, which leads to an acceleration of global warming. In addition to this, the release of freshwater into the ocean caused by the melting of glaciers can cause instability in ocean currents, which in turn affects the circulation of heat across the world. These feedbacks have the potential to exacerbate the effects that glacial retreat has on the climate system of the Earth.

5.4. Effects on the Different Weather Systems

It is possible for glacial retreat to have an effect on weather patterns, and this is especially true in areas where glaciers play an important part in the regulation of the local climate. Variations in temperature gradients, atmospheric circulation, and precipitation can all contribute to alterations in the patterns of the weather systems that are observed. These changes may lead to an increase in the frequency and severity of extreme weather events such as heatwaves, storms, and extended periods of drought.

5.5. International Consequences

The repercussions of glaciers melting away are not exclusive to the polar areas alone. The effects of climate change, which include rising sea levels, altered patterns of ocean circulation, and shifting weather patterns, are worldwide in scope. Coastal communities all throughout the world are at risk because of rising sea levels, and alterations to the weather systems can have an impact on agriculture, water resources, and economics. Climate refugees, people who have been forced to flee their homes because of the effects of melting glaciers and rising sea levels, present problems to migration and regional stability.

The melting of glaciers is one of the most obvious and striking signs of climate change on a global scale. The principal factor contributing to glacial retreat is global warming, which is caused by a rise in the amount of greenhouse gases present in the atmosphere of the Earth. The melting of glaciers causes freshwater to be released into the oceans, which contributes to an increase in sea level, disrupts ocean circulation, and has an effect on climate feedback mechanisms. The effects of glacial retreat are felt not only in polar regions but also across the rest of the world, having an effect not only on ecosystems but also on water resources, weather patterns, and human societies.

In order to understand the processes that are driving glacial retreat and the implications that it will have, scientific research and monitoring initiatives are essential. Researchers obtain insights into glacier behavior and anticipate future changes through the use of field observations, remote sensing, ice core analysis, glacier mass balance monitoring, and modeling. These initiatives assist educate climate policies and adaptation measures in order to overcome the issues faced by the melting of glaciers.

In order to address the problem of glacial retreat, a multidimensional approach is required. This approach must involve the reduction of drivers of global warming, the mitigation of greenhouse gas emissions, and the development of adaptive measures to deal with the effects of glacier melt. The ongoing transformation of our planet that is being caused by the receding of glaciers demonstrates the critical urgency of addressing climate change and protecting these important natural features.

4.1 Connection between glacial melt and freshwater resources

The existence of life, ecosystems, and human activities on our world are all dependent on the precious and limited resource that is freshwater. The availability of freshwater is becoming increasingly vital, and its management is becoming increasingly difficult, as the global population continues to expand in tandem with the acceleration of climate change. The water that is melted from glaciers is one of the most important sources of freshwater, and glaciers are intricately related to the freshwater resources of the world. In this essay, we will investigate the complex relationship that exists between glacial melt and freshwater resources. We will focus on the importance that these resources play in supplying clean water and in maintaining ecosystems, as well as the potential consequences that could result from glaciers receding as a result of climate change.

The Importance of Glacial Meltwater to the Nation's Supply of Freshwater

Making Available a Source of Drinking Water:

The melting of glaciers provides an important source of freshwater for many communities, particularly those located in high mountain regions and the polar regions. The glacial meltwater supplies these regions with all of their freshwater requirements, including that for drinking water, irrigation for agriculture, and the creation of electricity. Because glacier meltwater is often of a high quality and is devoid of pollutants and toxins, it is an invaluable resource for human use. Glaciers meltwater may be found in many places across the world.

Keeping the Streamflow Going:

The melting of glaciers has an effect on the flow of rivers and streams, which is especially noticeable during the dry season or during extended periods of drought. It functions as a natural buffer, discharging freshwater at the precise moment when it is required the most. This helps to minimize the consequences of water scarcity by ensuring a continuous supply of water for both ecosystems and human groups to use, which in turn helps to reduce the impact of water shortage.

Providing Maintenance for Ecosystems:

Glacial meltwater is essential to the survival of a great number of aquatic ecosystems, particularly those located downstream from mountain ranges. The temperature, sediment load, and nutrient content of glacier-fed rivers all combine to produce one-of-a-kind environments for the flora and fauna that live in these ecosystems. The absence of glacial meltwater can wreak havoc on these ecosystems, having a negative impact not just on the species diversity but also on the general health of aquatic life.

The Production of Hydroelectricity:

The meltwater from glaciers is an essential resource for the development of electricity. It does so by offering a steady supply of water to hydroelectric facilities, so contributing to the generation of electricity that is both clean and renewable. Hydropower is the principal source of electricity production in many nations, particularly those that have a substantial amount of glacial cover.

The Effects of Glacial Meltwater on the Water Supplies of Regional Areas

The Himalayan Mountains :

The Himalayan region is frequently referred to as the "Water Tower of Asia" due to its huge glaciers, which serve as an essential source of water for more than one billion people. Countries such as India, China, and Nepal are extremely reliant on the glacial meltwater that comes from this region for their agricultural and domestic needs as well as the generation of energy. These nations face a substantial risk to their water supply as a direct result of the melting of Himalayan glaciers, which is caused by climate change.

Mountains of the Andes :

Another significant source of glacier meltwater is found in the Andes Mountains, which are located in South America. Lima, the capital of Peru, and other cities in the country are dependent on the water resources that are provided by the glaciers in the Cordillera Blanca. The melting of these glaciers has already resulted in water shortages as well as disagreements over how water should be distributed.

Western regions of the United States:

Glacial melt contributes significantly, if not entirely, to the water supply of the Rocky Mountains and the Sierra Nevada range, both of which are located in the western United States. This water is essential not only for urban areas such as Los Angeles, Las Vegas, and Denver but also for agricultural regions. The melting of these glaciers is putting additional stress on water resources in the region, which are already being pushed to their limits.

Climate Change and the Melting of Glaciers

The rate at which glaciers are melting is accelerating as a result of climate change, which is upsetting the delicate balance that exists between glaciers and freshwater resources. The melting of more glacial ice and the retreat of glaciers are both caused by rising global temperatures, which has a number of significant repercussions, including the following:

Availability of Fresh Water Is Becoming Less Common:

The melting and disappearance of glaciers leads to a reduction in the amount of freshwater that is available for consumption by humans as well as by ecosystems. This can result in the loss of essential habitat for aquatic animals as well as a severe lack of water supply in areas that are highly dependent on the water that glaciers melt.

Changes Made to the Water's Quality:

The absence of glacial meltwater can also have an effect on the quality of the water. Because it often has a low amount of silt and other impurities, glacial meltwater is an extremely useful supply of pure water. There will be less glacier meltwater, which will likely result in water sources becoming more cloudy and possibly poisoned. This will have an impact on both humans and aquatic life.

Adaptations to Alterations in Streamflow Patterns:

Changes in streamflow patterns may result from a reduction in the amount of glacial meltwater.

The loss of glacial meltwater, which acts as a buffer during dry seasons, can make drought conditions worse, having an effect not just on agriculture but also on the dependability of water supply.

Increased Potential for Floods Caused by Glacial Lake Outbursts

When glaciers melt, they frequently leave behind glacial lakes that are kept from overflowing by natural dams formed by moraine. In the event that the dams break, which would cause massive volumes of water to be released farther downstream, these lakes could pose a considerable risk of glacial lake outburst floods (GLOFs). The rate of glacial retreat and the amount of accumulated meltwater in these lakes both contribute to an increased likelihood of GLOFs occurring.

Strategies for Adaptation as well as Mitigation

Management of Water Resources and Water Conservation:

The impacts of diminished glacial melt can be helped to some degree by taking preventative measures such as the development of effective water management techniques and the promotion of water conservation practices. This includes the development of better systems for storing, distributing, and recycling water.

Agriculture that is Environmentally Sound:

The use of sustainable agriculture practices has the potential to cut down on water usage as well as the reliance on meltwater from glaciers for irrigation.

Water conservation can be accomplished by the use of methods such as drip irrigation, crop rotation, and the planting of drought-resistant plant species.

Improvements in Hydropower Generation and Utilization:

It is possible to lessen reliance on meltwater from glaciers as a source of electricity generation by increasing the efficiency of hydropower plants and diversifying energy sources. Investigating other forms of renewable energy, such as solar and wind power, are two examples of the kinds of possibilities that can help alleviate the impact of losing this precious resource.

Reducing the Impacts of Climate Change:

It is essential to take action against the underlying cause of the problem, which is climate change, in order to lessen the associated hazards and lower the rate at which glaciers are melting. It is necessary to battle climate change and safeguard glaciers by putting into place policies that will cut greenhouse gas emissions and transition to an economy with lower levels of carbon emissions.

Systems for Monitoring and Providing Early Warning:

The potential risks that are connected with glacial retreat can be reduced by the development of monitoring systems and early warning measures for glacial lake outburst floods. Evacuation in a timely manner and being prepared for a disaster are two ways to preserve lives and limit property damage.

Cooperation on the International Level:

There are several glaciers and the resources that come from their meltwater that are shared between countries. It is absolutely necessary for countries that are geographically close together to cooperate in order to successfully manage and distribute these resources and to avoid conflicts.

The relationship that exists between the melting of glaciers and the availability of freshwater is an essential component of the world's water cycle. Glacial meltwater is an essential source of freshwater for many locations, ensuring the continued existence of human communities as well as ecosystems and the generation of electricity. However, the increased rate of glacial retreat that is a direct result of climate change creates enormous difficulties to the availability of freshwater, as well as to the quality of water and the stability of aquatic ecosystems.

It is vital to implement measures for adaptation and mitigation in order to meet these difficulties and guarantee the long-term viability of freshwater resources.

We can strive toward conserving the complicated connection between glacial melt and freshwater resources if we are efficient in the management of water resources, convert to sustainable agriculture methods, and reduce emissions of greenhouse gases. When we do this, we protect not only our access to clean water but also the delicate balance of ecosystems that are dependent on this important resource. This is a win-win situation.

4.2 Implications for tourism and local economies

The creation of jobs, the generating of money, and the improvement of infrastructure are all key aspects of economic development that are greatly influenced by tourism in a number of different places. The success of the tourist industry is intricately entwined with a variety of elements, including natural landscapes, cultural heritage, and local communities. tourist is frequently the principal source of income for local economies. However, this reliance on tourism also brings about issues and repercussions that have the potential to disrupt the communities' social, economic, and environmental fabric. In this in-depth study, we will delve into the multifaceted consequences for tourism and local economies. Specifically, we will investigate both the good contributions and the issues that occur as a result of the growth of the tourism industry.

The Potentially Favorable Effects on Tourism and the Economies of the Area

Economic Expansion and the Generation of New Jobs:

Tourism is a major contributor to economic expansion since it results in increased revenue for local establishments such as hotels, restaurants, and other service providers. The influx of tourists results in a demand for goods and services, which in turn results in the formation of new enterprises as well as prospects for employment. In many areas, tourism acts as an impetus for the formation of new businesses, both large and small, and contributes significantly to the vitality of the regional economy.

The Development of Infrastructure:

It is common for the requirement to accommodate tourists to be the driving force behind the construction of infrastructure, which may include networks of transportation, hotels, restaurants, and recreational facilities. Not only does this infrastructure improve the visitor experience, but it also helps the local population improve their quality of life and access to services that are necessary to their daily functioning.

Preservation of Culture and Cultural Exchange:

Through the facilitation of cultural interchange between tourists and local populations, tourism contributes to increased awareness of and respect for the world's myriad cultural traditions.

Tourists are frequently exposed to local customs, works of art, and handicrafts, which creates opportunity for the preservation of cultural heritage as well as the passing on of traditional knowledge from one generation to the next. Cultural tourism programs have the potential to benefit economically while also assisting communities in preserving their identities and histories.

Expanding the Scope of Economic Activity:

If a community's economy is dependent on just one sector, it may be more susceptible to cyclical shifts and unexpected events from the outside world. Communities can lessen their reliance on a particular industry by increasing their participation in tourism, which provides an opportunity for economic diversification. Diversification can help to establish a local economy that is more robust and sustainable, which in turn helps to reduce the risks that are connected with economic downturns in other sectors.

Community Amplification To be healthy:

If tourism is successful, local residents may see an improvement in their standard of living as well as an increase in their income. There is an opportunity to reinvest the money made from tourism-related activities in community improvement endeavors, such as those pertaining to education, healthcare, and social welfare. Residents of the area can, as a direct result, reap the benefits of enhanced infrastructure and social services, resulting in an overall improvement in their quality of life.

Problems and Unfavorable Consequences for Tourism and the Economy of Local Communities

Impact on the Environment Due to Excessive Tourism:

Overtourism, which is defined as an excessive influx of tourists that exceeds the capacity of a destination to accommodate them, can result in environmental deterioration, a strain on natural resources, and increased pollution. In major tourist areas, over crowding can cause damage to sensitive ecosystems, contribute to the accumulation of waste, and disrupt local biodiversity, all of which can lead to the long-term destruction of the environment.

Unstable employment due to factors such as seasonality

In many tourist destinations, the quantity of visitors might be unpredictable, which can result in employment patterns that are more seasonal. It may be difficult for local inhabitants who work in the tourism industry to maintain a steady livelihood throughout the year as a result of revenue changes brought on by seasonal shifts because the tourism industry is very susceptible to these shifts. This seasonal dependence might worsen existing income inequality and hamper the economy's ability to remain stable over the long term.

The commercialization of culture and the resulting loss of authenticity:

It is possible that the commercialization of local customs and cultural activities for the consumption of tourists may lead to the commodification of culture, which will ultimately result in the loss of authenticity as well as cultural identity. This phenomena has the potential to warp local customs and traditions, which may result in the cultural homogenization and the loss of values that are unique to a culture.

Local Communities That Have Been Displaced:

As tourism development continues, there is a risk that local residents will be uprooted as a consequence of the construction of tourist facilities and the rising cost of living. This can result in the disruption of social order within communities, as well as the loss of cultural practices and traditional means of subsistence. The gentrification of specific neighborhoods is another effect that can be caused by displacement. Gentrification makes it difficult for long-term inhabitants of a region to afford housing and to keep their existing social networks.

Revenue Losses Sustained from Tourism:

It is possible that a sizeable percentage of the profits made from tourism could be siphoned off and sent directly to the coffers of multinational firms or private investors located in other countries. As a result of this leakage, the advantages of tourism expenditure may not completely circulate within the community, which might reduce the multiplier effect that tourism has on the local economy. As a consequence of this, it's possible that residents and companies in the surrounding area won't completely benefit economically from the tourism industry.

Strategies for Mitigation and Management of Impacts

Methods of Environmentally Responsible Tourism:

The adoption of eco-friendly hotels, waste management systems, and ethical tourist rules are all examples of sustainable tourism practices that can help reduce the adverse effects of tourism on the natural environment. It is possible to help to the preservation of natural and cultural assets by urging tourists to respect the ecosystems and cultures of the area they are visiting.

Participation in Community Life and Fostering Autonomy:

It is possible to give local communities the power to actively participate in the development and management of tourism activities by including them in the planning and decision-making processes related to tourism and involving them in those processes. Community-based tourism programs, in which local inhabitants are directly involved in offering authentic cultural experiences to tourists, have the potential to generate a sense of ownership and pride, which in turn can lead to equitable distribution of the advantages of tourism.

Increasing the Variety of Tourist Attractions:

It is possible to assist spread the flow of tourists and decrease the load that is placed on popular sites by diversifying the tourism offers beyond the typical tourist attractions. Visitors can be encouraged to explore lesser-known sites by promoting off-the-beaten-path experiences, eco-tourism, and cultural immersion programs. This helps to support the development of developing tourism hubs and provides economic benefits to local economies in areas that are not the key tourist centers.

Building People's Capacity and Developing Their Skills:

It is possible to improve local workers' job opportunities and produce a workforce that is both more skilled and more competitive within the tourism industry by making investments in the education and skill development of the local workforce. Residents of a community can be equipped with the skills essential to deliver high-quality services and experiences for tourists by participating in training programs that focus on hospitality, language skills, and cultural understanding.

Regulations Regarding Tourism and Responsible Advertising:

Helping to regulate the expectations and behaviors of tourists can be facilitated by the implementation of responsible marketing tactics that encourage cultural sensitivity and sustainable tourism practices. Establishing and enforcing rules that protect the environment, preserve cultural heritage, and ensure fair labor practices can contribute to the sustainable expansion of the tourism industry while protecting the interests of local communities. These regulations should be created with the goals of protecting the environment, preserving cultural heritage, and ensuring fair labor practices.

Reinvestment of Proceeds from Tourism:

It is possible to ensure that the economic benefits of tourism directly contribute to the well-being of local residents by giving priority to the reinvestment of cash earned from tourism into projects aimed at community development, the enhancement of infrastructure, and the conservation of the natural environment. It is possible for local communities to be able to allocate cash from tourism to projects that fulfill their unique requirements and top priorities provided transparent financial systems and community-based funds are established.

The ramifications for tourism and local economies are varied, and in order to ensure sustainable development and equal distribution of benefits, a holistic approach is required. Tourism has the potential to be a substantial source of economic growth and cultural interchange; yet, the unchecked growth of the tourism industry can have negative effects on the surrounding people, the natural environment, and cultural heritage.

It is feasible to capitalize on the positive effects that tourism has while at the same time minimizing its negative effects if sustainable practices are put into place, community engagement is encouraged, and responsible tourism management is prioritized. When it comes to developing a vibrant tourism industry that contributes to the long-term well-being of both tourists and the communities in which they stay, striking a balance between economic development and the preservation of the environment and culture is essential.

Chapter 5

Thawing Permafrost

In recent years, thawing permafrost has received an increasing amount of attention due to the profound and far-reaching implications it has. This phenomena is a direct outcome of global climate change. The term "permafrost" refers to the frozen ground that can be found throughout huge parts of the Earth's polar and subpolar regions, such as the high-altitude mountain ranges, the Arctic, and the subarctic. Because of the warming of the planet's temperature, the permafrost is melting at a rate that has never been seen before. This results in the release of greenhouse gases, the disruption of ecosystems, and the creation of new difficulties for infrastructure, land use, and the environment. In the course of this in-depth investigation, we will investigate the factors that lead to the thawing of permafrost, as well as its effects and repercussions, as well as the climate feedbacks that may result, and we will investigate several ways for mitigating the adverse effects of this critical environmental problem.

Acquiring Knowledge about the Permafrost

Permafrost is described as soil or sediment that has a temperature that has remained below freezing (or 0 degrees Celsius or 32 degrees

Fahrenheit) for at least two years in a row. Due to the fact that it is made up of a combination of ice, dirt, organic materials, and minerals, the ecosystem there is frequently delicate and easily damaged. Depending on the magnitude and length of time that it remains in a frozen condition, permafrost can take on a number of distinct manifestations. These include continuous permafrost, discontinuous permafrost, and occasional permafrost.

1. **What Is Causing the Permafrost to Thaw?**
   1.1. An Increase in Temperature:
   The emission of greenhouse gases, particularly carbon dioxide (CO_2) and methane (CH_4), from human activities such as the burning of fossil fuels, deforestation, and industrial operations has been the primary contributor to the gradual but consistent rise in the average temperature of the Earth. The Arctic and subarctic regions are warming at a rate that is more than double the average for the rest of the world, which is causing permafrost to thaw.

   1.2. The Effect of Albedo:
   The albedo effect refers to the way in which snow and ice, which are both highly reflective, cover permafrost regions. This causes a considerable percentage of the incoming solar energy to be reflected back. As the permafrost thaws and exposes darker surfaces like soil and plant, the albedo effect declines. This leads to higher heat absorption and additional warming in a positive feedback loop because darker surfaces absorb more heat than lighter ones.

   1.3. Activity of Microorganisms:
   When permafrost thaws, the microorganisms that live beneath it become active and begin to degrade organic matter, which results in the emission of greenhouse gases such as carbon dioxide and methane into the atmosphere. One of the most important ways that greenhouse gases are released into the atmosphere is through the process of microbes breaking down organic carbon that is

frozen in permafrost.

1.4. Methane Hydrates:

Methane hydrates are a kind of methane that are trapped in ice and can be found in some areas of permafrost. These hydrates have the potential to release methane as the permafrost begins to thaw. Methane is a strong greenhouse gas that has a considerably larger capacity to trap heat than CO2 does over short durations.

2. **Permafrost Defrosting and Its Repercussions**

2.1. Emissions of Greenhouse Gases:

The melting of permafrost causes a release of carbon dioxide and methane, both of which add to the global concentration of greenhouse gases. This, in turn, has the effect of intensifying the planet's warming. This particular positive feedback loop, which is sometimes referred to as the permafrost carbon feedback, makes climate change even worse.

2.2. The Disruption of Ecosystems:

Ecosystems that thrive in permafrost are highly specialized and have evolved to circumstances of extreme cold. These ecosystems are being thrown off balance by the melting of the permafrost, and it may become difficult for native animals to adapt to the shifting conditions. The thawing of permafrost can also cause changes in flora, which can in turn have an effect on the populations of many wildlife species.

2.3. Subsidence of the Land:

The melting of permafrost can lead to land subsidence, which in turn can cause buildings, roads, and other forms of infrastructure to sink. Particularly in regions where there is a significant amount of permafrost, this might result in harm to towns, transportation networks, and energy infrastructure.

2.4. Erosion of Coastal Lands:

In areas of the coast that are abundant in permafrost, the melting of permafrost causes an acceleration in the pace of coastal erosion because the loss of frozen ground makes coastlines less stable.

This poses a huge risk to the towns that are located along these coastlines, as well as to the cultural and historical sites that may be found there.

2.5. The Quantity and Quality of the Water:

Alterations to the permafrost have the potential to influence both the quality and amount of the water. The thawing of permafrost can cause the discharge of pollutants and toxins into freshwater sources, which can have an impact on human health as well as the health of aquatic ecosystems. Alterations in hydrological regimes can also have an effect on the availability of water, particularly in areas of the world where permafrost acts as a water storage reservoir.

2.6. Impact on Culture and Society:

Indigenous groups located in the Arctic and subarctic regions are particularly susceptible to the effects of the thawing of permafrost because of their geographical location. It has the potential to weaken the cultural and social fabric of these communities, as well as disrupt customary practices, impact the availability of food resources, and affect the availability of food.

3. **Feedbacks on the Climate**

3.1. Carbon Feedback:

Carbon feedback happens when frozen permafrost melts and releases carbon dioxide and methane into the atmosphere in the form of CO_2 and CH_4, respectively. This, in turn, causes an increase in the concentration of greenhouse gases, which warms the temperature even more and speeds up the rate at which permafrost thaws, so generating a cycle that reinforces itself.

3.2. Feedback on the Albedo:

The thawing of permafrost, which exposes darker surfaces, has the effect of reducing the albedo, which in turn leads to an increase in the amount of solar radiation absorbed and higher temperatures. This results in a faster rate of thawing, which in turn perpetuates the cycle of climate change and the destruction of permafrost.

4. **Strategies and Answers for Damage Mitigation**

Emissions of Greenhouse Gases Should Be Cut:

It is imperative that there be a coordinated effort made to lower global emissions of greenhouse gases in order to impede the rate of permafrost thaw and the attendant effects. This involves the transition to energy sources that are cleaner and more sustainable, the improvement of energy efficiency, and the implementation of regulations to limit emissions from the industrial, transportation, and agricultural sectors.

Encourage the Responsible Use of Land:

Permafrost deterioration is something that can be mitigated with the use of sustainable land management methods that protect and preserve permafrost ecosystems. A few examples of these activities are responsible mining and resource exploitation, sustainable agriculture, and restricting urban expansion in places that are ecologically fragile.

Resilience of the Infrastructure:

It is vital to construct and maintain infrastructure that is capable of withstanding the challenges that are brought by the thawing of permafrost. This comprises the construction of buildings, pilings, and foundations that are capable of adapting to subsidence and coastal erosion. When building infrastructure projects, engineers and planners should take into account the one-of-a-kind qualities that are present in places with permafrost.

Observe and Do Research on:

Continuous monitoring and scientific research are very necessary in order to get a deeper understanding of the mechanisms involved in the thawing of permafrost, as well as its effects and the potential responses to them. This research can help communities and ecosystems make better decisions, provide more accurate risk assessments, and develop adaptive strategies.

Resilience of Communities and Adaptation to Change:

Indigenous and local populations living in areas with permafrost are obligated to devise adaptation techniques that will assist them in adjusting to the shifting environmental conditions. These solutions may involve the use of traditional knowledge, infrastructure that is adaptable to climate change, and diverse livelihoods in order to lessen dependency on ecosystems that are susceptible to damage.

Collaboration across International Boundaries:

The thawing of permafrost is a worldwide problem that demands cooperation and coordination on an international scale. The sharing of information, research, and resources can help in the process of developing methods to effectively address this situation. The Paris Agreement, the primary goal of which is to reduce the effects of climate change, is a crucial framework for tackling challenges that are associated with permafrost.

Restoration of the Permafrost:

The reestablishment of permafrost areas and the strengthening of their resistance may be part of the restoration process. In order to restore permafrost, several methods such as the management of vegetation and the stability of soil to prevent erosion are utilized. Additionally, efforts are made to both maintain cold permafrost areas and encourage regeneration.

The thawing of permafrost is a complicated and multi-faceted problem that overlaps with climate change, environmental degradation, disruption of ecosystems, and the effects of society. The permafrost thaw is gaining momentum, which poses a huge risk to the future of our planet as average temperatures throughout the world continue to climb. It is essential to get an understanding of the causes, implications, and potential solutions for the thawing of permafrost in order to lessen the negative effects of this process and build resilience in communities and ecosystems that are now at risk.

It is crucial to realize that the management of permafrost is strongly tied to larger efforts to battle climate change. Even if tackling the issue of thawing permafrost is a complicated and continuing process, it is essential to recognize this connection. Reducing emissions of greenhouse gases and making the transition to a future that is more sustainable and resilient are both essential actions that must be taken in order to address the thawing of permafrost and the far-reaching ramifications of this process. Furthermore, finding effective answers to this urgent global crisis requires a combination of international cooperation, scientific research, and the active involvement of local communities.

5.1 Understanding permafrost and its role in carbon storage

There is much more to the study of permafrost than meets the eye when you hear the word "permafrost," which conjures up pictures of frozen landscapes and intense cold. It is extremely important for the sequestration of carbon, the control of climate, and the maintenance of the fragile equilibrium of the ecosystems on our planet. The term "permafrost" refers to the permanently frozen soil or sediment that can be found in polar and high-altitude regions. This type of soil or sediment is known to contain significant quantities of organic carbon. This carbon has been stored for thousands of years, but due to recent climate change, it is now in danger of being destroyed.

In this investigation of one thousand words' length, we will delve into the fascinating world of permafrost, its function as a repository for carbon, and the possible repercussions of its melting.

What exactly is the permafrost?

As its name suggests, permafrost is ground or sediment that is permanently frozen over the course of an entire year. It is defined as ground that has a temperature that remains below freezing (0 degrees Celsius or 32 degrees Fahrenheit) for at least two years in a row. Even though the Arctic and Antarctic regions are where permafrost is most usually found, it may also be found in high-altitude places such as the Tibetan Plateau and even in some alpine regions. This may come as a surprise to some people. The depth of the permafrost can range anywhere from just a few centimeters to several hundred meters, therefore its thickness can change dramatically.

The components that make up permafrost are soil, gravel, and rock; it also has ice, which is

what holds all of these components together. It is possible for there to be as much as ninety percent of the volume of the permafrost that is composed of ice. The characteristic stiffness and strength of permafrost is due to the presence of an icy matrix.

Carbon that is found in permafrost

The amount of carbon that may be found in permafrost is one of the most fascinating characteristics of this natural phenomenon. The organic carbon found in permafrost comes in the form of decomposed plant material and other forms of organic matter. This carbon has been there for thousands of years and has accumulated in the permafrost. Because of the low temperatures, this organic carbon has been kept in good condition because microbes find it difficult to decompose it. In practice, permafrost performs the function of a natural freezer, storing large quantities of carbon underneath the frozen surface of the permafrost.

The quantity of carbon that is expected to be trapped in permafrost all across the earth is astonishing. It is expected to contain more than twice as much carbon as can be found in the atmosphere of the Earth at the present time. This massive store of carbon exists in the form of frozen organic debris, which may contain everything from fossilized trees to microbial remains and dead animals. This carbon will be effectively sequestered for as long as the permafrost continues to remain frozen. This will prevent it from becoming a part of the carbon cycle and contributing to emissions of greenhouse gases.

Climate Change and the Melting of the Permafrost

As a result of global warming, the long-term integrity of permafrost is currently being put in jeopardy. The permafrost begins to thaw as temperatures rise, releasing the carbon that it has been storing for decades or perhaps centuries.

Because it has the potential to intensify the effects of climate change, this thawing process is a key worry for climate scientists and policy-makers.

When permafrost thaws, the organic stuff that is contained within it is made accessible to the activities of microbes. The once-frozen carbon then begins to be broken down by microbes, which results in the production of carbon dioxide (CO2) and methane (CH4), both of which are powerful greenhouse gases. Methane in particular is a matter for concern since, over shorter periods of time, its potential to act as a heat trap is significantly larger than that of carbon dioxide.

A positive feedback loop is created when these greenhouse gasses are released into the atmosphere as a result of thawing permafrost. The release of additional carbon into the atmosphere has the effect of contributing to global warming, which in turn hastens the process of permafrost thawing. This virtuous loop can have far-reaching effects on our planet's climate if it is allowed to continue.

Influence on the Changing Climate

The possible effects that melting permafrost could have on climate change are a source of significant worry. Through research and simulation, scientists have been attempting to better understand and quantify the consequences of this phenomenon. Even while it is difficult to forecast the exact extent of permafrost melt and the carbon release that is connected with it, there are a few essential points that are known.

To begin, the release of carbon from melting permafrost has the potential to considerably increase the quantity of greenhouse gases in the atmosphere. According to a number of studies, the thawing of permafrost could add anywhere from 70 to 130 gigatons of carbon to the atmosphere by the year 2100. This estimate is dependent on the amount of warming that will occur in the future. This is a significant amount, which is equal to the emissions that have been produced by all human activity in the United States over the course of ten years.

Second, the melting of permafrost could result in the release of methane, which is a powerful greenhouse gas that could cause an increase in the rate of current warming. Over a period of 100 years, the ability of methane to retain heat in the atmosphere is over 25 times more effective than that of carbon dioxide. Although methane has a shorter lifetime in the atmosphere than CO2, its initial influence is significantly more significant.

Third, the effects of the thawing of permafrost are not confined to the regions of the Arctic and Antarctica by any means. Carbon and methane that have been released can enter the global circulation system, which can contribute to heat and climate change on a global scale.

The Value of Investigating Matters Further

The complex scientific problem of trying to understand the dynamics of permafrost melt and its involvement in climate change is now being worked on. For their research on permafrost, scientists use a wide variety of approaches, including measurements taken on the ground, satellite photos, and computer simulations. Using these methodologies, scientists are better able to evaluate the magnitude of the permafrost melt, as well as its rate and the prospective carbon emissions.

It is essential to do research on permafrost in order to improve climate models and accurately estimate the future effects of thawing. Enhanced models can help inform decisions about public policy and lead attempts to reduce the effects of climate change. In addition, we are able to build methods to adapt to the changes that are already taking place as a result of this research.

Adjusting to the Thawing of the Permafrost

Damage to Infrastructure The thawing of permafrost can cause buildings, roads, and other types of infrastructure to sink and become unstable. This can be a dangerous situation.

Erosion: The thawing of permafrost can lead to an increase in coastal erosion, which puts towns and natural areas in jeopardy.

Alterations in Water Availability The thawing of permafrost can cause hydrology to change, which in turn can have an effect on the availability and quality of freshwater supplies.

Disruption of Ecosystems The shifting environmental circumstances have the potential to disrupt local ecosystems, as well as the people that rely on those ecosystems for their source of food and other resources.

Adaptation techniques are currently being developed in order to meet these issues. Some examples of these efforts include developing structures with foundations that are compatible with permafrost, applying erosion control measures, and managing water resources more effectively. Indigenous populations, which have been living in permafrost regions for decades, frequently have rich traditional knowledge that can inform adaptation efforts.

Inhibiting the Thawing of Permafrost

The larger attempt to slow climate change has a close relationship with the more specific goal of slowing the thawing of permafrost. The reduction of greenhouse gas emissions, in particular those resulting from the combustion of fossil fuels, is the key technique for reducing the effects of permafrost thaw. We can delay the rate of permafrost thaw and lower the release of carbon and methane into the atmosphere if we take steps to control global warming.

Planting trees and other types of vegetation in areas of permafrost can help cool the soil and lessen the rate at which it thaws. This practice is known as reforestation. Trees not only give shade but also help to stabilize the soil, both of which contribute to a slower rate of carbon release.

Albedo Enhancement: Decreasing heat absorption and delaying the thawing of permafrost can be accomplished by increasing the surface's reflectivity. This can be accomplished, for example, by painting buildings and roads in lighter colors.

Carbon Sequestration: The process of developing technologies to capture and store carbon dioxide from the atmosphere is known as carbon sequestration. Carbon sequestration can be used to counteract emissions caused by the thawing of permafrost.

Capturing Methane: Technologies that can capture and make use of the methane that is emitted as a result of thawing permafrost can help decrease the influence that this process has on the climate.

The process of reducing the rate at which permafrost thaws is complicated and difficult, but it is a crucial part of our attempts to curb the effects of climate change.

It is a common misconception that permafrost is a feature of the Earth's terrain that is permanently frozen and unchanging. It is a dynamic and essential component of the climate system of the globe, playing a vital role in the storage of carbon and the management of emissions of greenhouse gases. The future of permafrost is uncertain as the temperature of the earth continues to rise, and the ramifications of its thawing could have significant repercussions.

Both scientists and politicians place a high focus on expanding their knowledge of permafrost and its function as a carbon sink. Ongoing research is being conducted on this subject with the purpose of enhancing our understanding of the dynamics of permafrost and enhancing our capacity to both predict and mitigate the consequences that it has on the climate.

A multi-pronged strategy is required in order to effectively address the issue of permafrost thaw. We need to investigate new approaches to directly address the issue of permafrost melt, develop strategies to adapt to the changes that are currently taking place, and cut emissions of greenhouse gases in order to slow down or stop global warming. We can make strides toward preserving the frozen carbon reservoirs in permafrost and minimizing the potentially catastrophic repercussions of its release on our climate and our world if we take action on these various fronts.

5.2 Release of greenhouse gases from thawing permafrost

The phrase "permafrost," which conjures up an image of a frozen and unchanging environment, is not at all what it seems to be. It is a large carbon sink, holding vast quantities of organic material in a frozen state for millennia, which plays an important part in the global climate system and is one of the most important roles that it plays. However, this seeming state of stability is currently in jeopardy as a result of changes in the environment on a worldwide scale. The permafrost in the Arctic and other freezing locations is melting, which is releasing long-buried greenhouse gasses into the atmosphere. This process is being driven by rising temperatures. This process offers a significant risk to the global climate system, which is frequently underappreciated by scientists. In this piece, we will investigate the release of greenhouse gasses that occurs as a result of thawing permafrost, as well as its likely causes, implications, and potential solutions for mitigation.

The Elements That Make Up Permafrost

The definition of permafrost is ground that has a temperature that has remained at or below freezing for at least two years in a row. It is a massive carbon reservoir that is frequently ignored despite the fact that it spans about 24 percent of the land surface in the northern hemisphere. Over the course of thousands of years, organic matter such as decomposing plants and animals has been accumulating within this soil and silt that has been frozen solid. Because of the low temperatures, this organic matter has been kept; as a result, it has not undergone the decomposition that would have resulted in the emission of carbon dioxide (CO_2) and methane (CH_4) into the atmosphere.

Permafrost Defrosting and the Release of Greenhouse Gases

Warming of the planet's atmosphere is the principal factor contributing to the melting of permafrost. The permafrost in the Arctic and other frigid locations is becoming more susceptible to thawing as the average temperature of the planet continues to climb. This process of thawing can take place as a slow warming that occurs over a period of several decades or as a quick event that is driven by disturbances such as wildfires. Once the permafrost begins to thaw, the once-frozen organic material will become susceptible to the breakdown processes of microbes. Microbes are responsible for the breakdown of organic materials, which results in the production of CO_2 and CH_4. After that, these gases are released into the atmosphere, where they add to the greenhouse effect and make the problem of global warming even worse.

The powerful greenhouse gas known as methane

When talking about the thawing of permafrost, methane should be of special concern. Although carbon dioxide (CO2) is the more well-known greenhouse gas, methane's capacity to trap heat is far greater than that of CO2. When compared to carbon dioxide, the ability of methane to retain heat over a period of 20 years is around 84-87 times greater. The release of methane from thawing permafrost has the potential to greatly increase global warming in the short term, which might have disastrous effects on the environment.

Estimates in Light of the Uncertainty

Variations in both the total quantity of carbon that is believed to be stored in permafrost and the possible release of greenhouse gases can be attributed, in large part, to the uncertainties that surround permafrost research. According to a number of different estimations, the amount of carbon that is found in permafrost is around 1,400 to 1,600 gigatons, which is almost twice as much carbon as is now found in the atmosphere. If it were to be released, this enormous carbon storage has the potential to significantly magnify the effects of global warming.

According to projections made by the Intergovernmental Panel on Climate Change (IPCC), under a high-emission scenario, the average global temperature might rise by 1.5 degrees Celsius by 2030 and 2.5 degrees Celsius by 2040. The permafrost will thaw to a greater extent as a result of these temperature rises, which will then result in releases of greenhouse gases. These estimates, however, should not be taken too seriously because the possible releases might be anything from a few hundred to several thousand gigatons of carbon. The sheer degree of diversity exhibited by these estimations highlights the intricacy of the problem involving permafrost.

Iterative Loops of Feedback and Accelerated Thaw

The possibility of feedback loops that could speed up the thawing of permafrost and the release of greenhouse gases is one of the most significant causes for concern. For instance, the melting of permafrost can cause the surface of the ground to sink, so producing depressions that can become sources of water accumulation. These depressions may then be used as wetland areas, which would encourage the growth of bacteria that produce methane and lead to an even greater increase in methane emissions. These kinds of positive feedback loops have the potential to magnify the effect that the thawing of permafrost has on the climate system.

Permafrost Defrosting and Its Repercussions

Acceleration of Global Warming: The increased methane and CO2 emissions that will result from the thawing of permafrost will further intensify global warming, making it even more difficult to contain temperature increases to levels that are safe.

Ecosystems That Have Been Changed Melting permafrost has the potential to trigger subsidence, which can destabilize infrastructure and have an effect on the habitats of species.

Erosion of Coastal Lands: In many Arctic and sub-Arctic regions, the melting of permafrost is a contributing factor in the erosion of coastal land, which has an impact on communities and indigenous groups.

The thawing of permafrost can also result in the release of ancient microorganisms and other potentially dangerous pathogens, which poses a risk to human health.

Ocean Acidification The thawing of permafrost can lead to the release of carbon into the ocean, which in turn contributes to ocean acidification and has additional negative effects on marine species.

Strategies for Risk Minimization

Both the reduction of emissions caused by human activities and the implementation of measures to protect and restore the ecosystems of permafrost regions are necessary components of the efforts to minimize the release of greenhouse gases caused by the thawing of permafrost.

Emissions of Fossil Fuels Should Be Decreased The single most efficient strategy for reducing the rate of global warming and mitigating the effects of permafrost melt is to lower the amount of greenhouse gases emitted by human activities, particularly the combustion of fossil fuels.

Protecting permafrost regions from human disturbances such as deforestation and the construction of infrastructure can go a long way toward preserving the permafrost's stable state. One way to do this is by promoting sustainable land use.

Restoring Wetlands The process of restoring wetlands can help reduce methane emissions by encouraging the growth of bacteria that consume methane. This helps to counteract the amount of methane that is released as permafrost thaws.

Capturing and Storing Carbon: Research into methods for capturing and storing carbon may provide possibilities for catching and sequestering greenhouse gases released as a result of thawing permafrost.

Indigenous Knowledge and Collaboration: Working together with indigenous groups, whose members have spent generations living in permafrost zones, can be an effective way to incorporate traditional knowledge into climate change adaptation measures.

The emanation of greenhouse gases as a result of the thawing of permafrost is an essential and difficult component of the ongoing dilemma regarding the planet's climate. Its repercussions are not limited to the Arctic region and will have an effect on the entire world. In order to solve this issue, a multi-pronged approach is required. This approach must include initiatives to conserve and restore habitats that contain permafrost as well as large reductions in greenhouse gas emissions that are caused by human activity. As we work to lessen the impact of this impending danger, it is essential that we continue to improve our knowledge of the dynamics of permafrost and the role that it plays in the climate system of the Earth so that we may devise remedies that are both effective and founded in scientific evidence. If we are unable to do so, the climate and ecosystems of our world may be negatively impacted in ways that are more substantial and difficult to reverse.

5.3 Consequences for infrastructure, especially in the Arctic

The rapidly shifting environment in the Arctic, which is marked by rising temperatures, melting permafrost, and shrinking sea ice, has ushered in a new era of challenges for the development and maintenance of infrastructure. As a result of the retreating ice cap, there is a greater emphasis being placed on the construction of essential infrastructure for resource extraction, transportation, and research. This is because the region is becoming more accessible. However, these changes will have a wide range of repercussions, including the disruption of existing stability and an increase in the dangers faced by both native inhabitants and wildlife. This article investigates the intricate network of problems that threaten the infrastructure in the Arctic region, diving into the ramifications for both ongoing projects and those that are yet to be completed.

Instability of the Permafrost and Effects on Infrastructure

The thawing of the permanently frozen ground known as permafrost, which serves as the basis for many buildings and other structures located in the Arctic, is one of the key sources of concern over the region's infrastructure. The thawing of the permafrost that results from an increase in temperature makes the ground unstable. This, in turn, can cause damage to infrastructure as well as subsidence and ground settlement. Buildings, roads, pipelines, and other essential pieces of infrastructure are all at risk of suffering structural deterioration, which would render them hazardous and necessitate either expensive repairs or complete rebuilding. As the conditions of the permafrost continue to shift, there is an ever-increasing demand for novel engineering solutions that can accommodate these shifts.

Erosion of Coastal Lands and the Vulnerability of Infrastructure

Increasing temperatures and a reduction in the amount of sea ice have both contributed to an acceleration of coastline erosion in the Arctic, which further endangers the infrastructure that is already there. Coastal communities, including indigenous groups, are especially susceptible to the effects of erosion. This is due to the fact that key infrastructure, such as homes, ports, and other essential amenities, are at risk of being washed away by erosion. It is now an urgent need to relocate entire communities, which calls for extensive planning and the allocation of resources in order to preserve the social fabric and cultural history of these communities.

Problems confronting the Transportation and Shipping Systems

The reduction in the amount of sea ice in the Arctic has resulted in a rise in the number of maritime operations, such as commerce and the exploration of resources. Even if this creates brand new potential for the economy, it also creates issues for the infrastructure, notably with regard to the creation and maintenance of dependable transportation routes and ports. It is now very necessary to have proper port infrastructure, icebreakers, and navigational aids in order to guarantee the secure and effective circulation of products and resources. To ensure the continued growth of the region's economy in the face of unpredictability in ice conditions and shifting patterns of precipitation, it is essential to make investments in infrastructure that is both resilient and flexible.

Infrastructure for energy production and environmental sustainability

The Arctic region contains a significant amount of energy resources, such as oil, gas, and the possibility for renewable energy sources. The cold temperature and the constraints posed by melting permafrost make the mining and delivery of these minerals more difficult. However, these resources are still valuable. When the ground is unstable, oil and gas pipelines run the risk of being damaged. On the other hand, renewable energy projects confront issues when it comes to maintaining the ground's stability for wind turbines and solar panels. It is essential to strike a balance between the development of renewable energy sources and efforts to reduce the negative influence on the environment in order to protect the delicate ecosystem that exists in the Arctic.

The Effects on Native American Communities

The lack of adequate infrastructure in the Arctic has a significant negative effect on the indigenous groups that have lived off the land and the resources it provides for many generations. The disruption of key services, including transportation and communication, as well as other essential functions, has a direct impact on their way of life and their access to vital resources. The loss of traditional hunting and fishing grounds as a result of changes in the environment can have a substantial impact on the ways in which they make their living as well as their cultural traditions. It is vital to approach the planning and development of infrastructure in a way that is both inclusive and community-centered if one want to secure the preservation of indigenous cultures and ways of life.

Protecting the Flora, Fauna, and Ecosystems of the Arctic

The expansion of infrastructure in the Arctic poses a threat not only to the one-of-a-kind and delicate ecosystems that the region is home to but also to the animals that lives there. The natural habitat may be disrupted and put at risk as a result of activities such as construction, an increase in the number of humans present, and the possibility of oil spills caused by shipping and drilling operations. It is absolutely necessary to put into action rigorous environmental legislation, carry out comprehensive environmental impact studies, and make investments in environmentally responsible infrastructure activities in order to reduce the negative effects on the biodiversity of the Arctic.

Strategies for Mitigation as well as Adaptation

Enhanced Engineering Solutions It is essential to develop forward-thinking engineering practices that take into account the stability of permafrost and the erosion of coastal areas in order to guarantee the lifetime and resiliency of infrastructure in the Arctic.

Community Engagement and Empowerment: Involving local communities, particularly indigenous populations, in the decision-making process for the construction of infrastructure can assist to generate sustainable and culturally sensitive solutions. Involving local communities in the decision-making process can also help to empower those groups.

Environmental Safeguards: If long-term sustainability is to be achieved, it is imperative that strict environmental laws and monitoring procedures be put into place. These will help to minimize the impact that the expansion of infrastructure will have on the delicate Arctic ecology.

Investing in Research and Development It is essential to invest in research and development in order to create resilience in the face of a changing climate. This can be accomplished by advancing sustainable infrastructure technologies and practices that are adapted to the specific difficulties that the Arctic presents.

Facilitating International Cooperation and Collaboration Among Arctic states and Stakeholders It is vital to facilitate international cooperation and collaboration among Arctic states and stakeholders in order to address common infrastructural concerns and assure the region's sustainable development.

Because of the deep and complex effects that climate change will have on Arctic infrastructure, it will be necessary to take a strategy that is both comprehensive and collaborative if we are to meet the challenges faced by the rapidly shifting environment.

Stakeholders can strive toward the goal of constructing resilient infrastructure that strikes a balance between the advancement of economic activity and the conservation of the natural environment if they give sustainable practices top priority, encourage community engagement, and incorporate indigenous knowledge. In order to successfully navigate the intricacies of infrastructure development in this one-of-a-kind and fragile environment, it will be essential to embrace adaptive tactics and invest in new solutions. This will protect the region's future for future generations.

6

Chapter 6

Rising Sea Levels

The rise in sea levels is one of the most distinctive characteristics of our climate's current state of flux. Because of the ongoing warming of the planet's climate, the world's oceans are growing larger, and the ice that covers glaciers and the polar areas is melting at a more rapid rate. These processes are contributing to rising sea levels, which will have profound and far-reaching repercussions for the economies, ecosystems, and communities that are located along the shore. This article investigates the factors that are contributing to climate change's most significant symptom, which is rising sea levels, as well as the repercussions of these factors and some possible responses to the problems they provide.

1. **The Factors That Contribute to Rising Sea Levels**
   The expansion caused by heat
   The phenomenon known as thermal expansion is one of the key factors that contribute to the rise in sea level. The oceans are able to take in more heat and become more expansive as a direct result of an increase in the amount of greenhouse gases that are present

in the atmosphere. This thermal expansion causes the volume of the ocean to increase, which directly contributes to higher sea levels as a result of rising temperatures. It is believed that thermal expansion is accountable for around half of the rise in sea level that has been witnessed over the course of the past century.

Ice Sheets and Glaciers on the Verge of Melting

Another important contributor to the problem of increasing sea levels is the disintegration of ice sheets in polar regions and on mountain glaciers. The ice sheets that cover Greenland and Antarctica are suffering worrisome losses in mass as a direct result of the ongoing rise in global temperature. The meltwater that originates from these ice sheets and glaciers ultimately finds its way into the oceans, where it contributes to the overall amount of seawater. Recent research indicates that the polar ice sheets are shedding ice at a rate that is accelerating, which is contributing to a rate of sea level rise that is also rising.

The dissolution of the Arctic Sea Ice

Even while the melting of Arctic sea ice does not directly contribute to the rise in sea level because the ice is already in the ocean, the ramifications of this phenomenon have far-reaching effects. The melting of Arctic sea ice wreaks havoc on both local and global climate systems, causing shifts in ocean circulation patterns and contributing to an accelerated rate of warming. These changes can have an indirect influence on sea levels because they can affect the distribution of heat and currents in the seas, which makes them a contributing component to the rise in sea level.

Activities Involving Humans

The increase in greenhouse gas emissions that is the direct cause of global warming is being driven mostly by human activity, in particular the burning of fossil fuels, deforestation, and changes in land use. This, in turn, is the cause of many of the changes that are contributing to the rise in sea level. Additionally, the expansion of coastal areas and the extraction of groundwater in coastal areas can both lead to land subsidence, which effectively elevates sea levels relative to the land and increases the risk of coastal communities being flooded.

2. **The Repercussions of Increasing Sea Levels**

Erosion of the Coastline

Coastal erosion is the most obvious and immediate effect that comes as a result of rising sea levels. Waves and storm surges will be able to go deeper inland as sea levels continue to rise, severely eroding shorelines and consuming land. This process has the potential to result in the damage of infrastructure, including roads, buildings, and utilities, as well as the loss of valuable coastal property. It can also result in the displacement of communities. Erosion of coastlines is a major concern whenever there is a high population density along the coast or on islands with a low elevation.

A rise in the level of flooding

A correlation can be shown between rising sea levels and an increase in the frequency and severity of coastal flooding. Even even small rises in sea level can result in storm surges that are both more frequent and more damaging when they occur during extreme weather events. This can lead to widespread flooding in coastal cities and regions. Large metropolitan areas, such as New York City, Miami, and Shanghai, as well as smaller coastal villages and islands, are all at risk for flooding.

The Uprooting of Whole Communities

Low-lying coastal settlements run the risk of being inundated by rising sea levels, rendering them uninhabitable. As a result of increasing sea levels, the lives of millions of people all over the world are in danger of being uprooted. As they look for new homes and ways to make a living, climate refugees frequently face difficulties on the social, economic, and political fronts. Indigenous groups, in particular, are disproportionately affected as a result of the danger posed to the lands they have traditionally lived on and the forms of life they have practiced.

Infiltration of Saltwater

The infiltration of salt water into freshwater sources in coastal areas is a potential consequence of rising sea levels. This poses a huge risk to the agricultural industry since groundwater that has been contaminated by seawater might leave farms unfit for crop production. It also has an effect on the supply of drinking water, making it more difficult and expensive to deliver clean water to towns located along the coast.

Influence on the World's Ecosystems

Coastal ecosystems, such as wetlands, mangroves, and estuaries, are especially vulnerable to fluctuations in sea level because of their proximity to the water. These essential ecosystems, which serve as nurseries for marine life, provide protection from flooding, and support a vast array of wildlife, are at risk of being submerged and destroyed by rising sea levels. The extinction of these ecosystems has the potential to destabilize fisheries, worsen water quality, and make coastal areas less resistant to the effects of natural disasters.

Consequences for the Economy

The rise in sea level may have significant repercussions for the economy. There is a potential for enormous financial loss for individuals, businesses, and governments as a consequence of the destruction of infrastructure, the loss of property, and the increased danger of flooding. In addition, tourism, which is a significant economic driver in many coastal locations, is badly impacted by both the rise in sea level and the attendant repercussions of that rise.

3. **Suggestions for Alternatives and Adaptive Methods**

Efforts Made to Reduce Emissions of Greenhouse Gases

It is vital to reduce greenhouse gas emissions by actions such as transitioning to clean energy sources, boosting energy efficiency, and replanting in order to slow down the rate of global warming and, as a result, the rise in sea level. This objective has been significantly advanced by the signing of the Paris Agreement, which is a multilateral convention aimed at reducing the effects of global warming.

Planning for Adaptations

In order to address both the short-term and the long-term effects of increasing sea levels, coastal towns and governments need to engage in extensive adaptation planning. This includes taking precautions such as constructing or modifying infrastructure so that it can resist floods, implementing early warning systems, and enacting zoning restrictions that restrict development in regions that are susceptible to flooding.

Protection of the Coast

It is possible to lessen the effects of coastal erosion and floods by making investments in coastal protection measures. Some of these methods include seawalls, levees, and the restoration of sand dunes. These constructed solutions may bring some relief in the short term, but they should be considered part of a larger strategy that also includes natural solutions such as the restoration of mangrove forests.

Planning for the Sustainable Use of Land and Urban Space

The vulnerability of coastal regions can be reduced by the use of sustainable land use practices and urban design. This involves the preservation of green spaces, wetlands, and buffer zones that have the capacity to absorb surplus water during periods of precipitation. It is also essential to construct physical infrastructure that takes into account future rises in sea level, such as roads and buildings that are raised.

Regression and Repositioning

Communities that are at a high risk due to rising sea levels may find that retreat and relocation are the most practical options for them in some circumstances. This necessitates the relocation of residents away from dangerous coastal areas and into more secure areas. To ensure the well-being of communities that have been uprooted, rigorous planning and support are both necessary.

Cooperative efforts on a global scale

The problem of rising sea levels is one that affects the entire world, and in order to successfully solve it, international cooperation is required. The United states Framework Convention on Climate Change (UNFCCC) and the Intergovernmental Panel on Climate Change (IPCC) provide a platform for states to collaborate on mitigating the effects of global warming and developing measures to adapt to those effects.

Investigation and Observation

Understanding the developing difficulties and determining how well adaption methods are working requires ongoing research and monitoring of the rise in sea level and its repercussions. This knowledge can help shape public policy, provide support for decision-making, and raise awareness among the general public.

One of the most obvious and noticeable effects of climate change is the gradual but steady elevation of sea levels. The world's oceans are progressively expanding inland, which is having a negative impact on coastal people, ecosystems, and economies. This is happening for a variety of complicated reasons. The rise in sea level has far-reaching implications, some of which include the erosion of coastal areas, an increase in the frequency and severity of flooding, the relocation of towns, and the devastation of key ecosystems. Despite this, the global community does not stand helpless in the face of this obstacle. It is absolutely necessary to implement both mitigation initiatives, which aim to lower emissions of greenhouse gases, and adaptation plans, which aim to safeguard and prepare coastal areas. In order to solve this global issue and secure a more sustainable future for coastal regions and the world as a whole, international collaboration and proactive planning are vital components of the solution. The need for prompt and comprehensive action has made the issue of rising sea levels a defining one for both our age and the generations that will come after us.

6.1 Discussion on sea level rise and its measurements

The rise in sea level, which is a visible result of global warming and climate change, is an essential indicator of the way in which the climatic system of the Earth is changing. It is a consequence of the thermal expansion of saltwater as well as the melting of glaciers and ice caps in the polar regions, and it contributes to the flooding of coastal areas all over the world. Understanding the methodologies used to measure changes in sea level, the factors that influence these measurements, and the consequences for the global climate change assessment are vital if one is to design effective strategies for mitigating and adapting to the effects of rising sea levels and accurately assess the impacts of rising sea levels. This article presents a comprehensive discussion of the topic of rising sea levels, its measurements, and the broader consequences for climate science and policy.

1. **Comprehending the Raise in Sea Level**
   Reasons Behind the Increasing Sea Level
   The primary cause of an increase in sea level is global warming, which results in the thermal expansion of seawater as well as the melting of glaciers and ice sheets. The acceleration of polar ice melting and the resultant rise in global sea levels are both direct results of the increase in greenhouse gas emissions that have been caused, in large part, by human activities. These emissions have had the effect of intensifying the warming of the Earth's atmosphere. Alterations in the patterns of ocean circulation, such as the sluggishness of the Atlantic Meridional Overturning Circulation, are another factor that may have a role in the rise or fall of regional sea levels.

   Variability in the Height of the Sea Level
   The rise in sea level is not consistent all over the world; localized differences are caused by a number of reasons, including ocean currents, the subsidence of land, and the gravitational impacts of shifting ice masses. For instance, some coastal places suffer faster rates of sea level increase due to local geological processes, while others may even notice a temporary reduction in sea levels due to natural climatic oscillations like El Nio. This variation in sea level rise can be attributed to a variety of factors, including

global warming and natural climate oscillations. For the purpose of determining how rising sea levels may affect vulnerable coastal communities and ecosystems, it is essential to have a thorough understanding of these regional variances.

2. **Methods Used to Measure the Level of the Sea**
A Measurement of the Tides

The use of tide gauges, which are among the oldest and most conventional methods for measuring changes in sea level, is widespread. These instruments keep track of water levels in relation to a stationary location on land, which enables them to provide continuous, long-term data on tidal changes and extreme events. Tide gauges, despite the fact that they provide useful historical data, are limited in their ability to reliably reflect changes in global sea level because of their localized character.

Altimetry obtained from satellites

Satellite altimetry is a sophisticated method of remote sensing that monitors changes in sea level by carefully monitoring the height of the ocean surface from space. The technology is also known as "altimetrism." This method offers complete and precise data on changes in sea level throughout time, as well as a worldwide perspective on the variations in sea level that have occurred over time. The use of satellite altimetry has become an essential tool for both climate scientists and policymakers as it has led to substantial advancements in our understanding of the complexity of sea level rise.

The science of gravimetry

Gravimetry is the practice of monitoring the gravitational field of the Earth to discover variations in mass distribution. These variations might include shifts in the distribution of ice mass and water. Researchers are able to calculate the percentage of total sea level rise that can be attributed to melting ice sheets and glaciers by examining gravitational anomalies. When paired with other methods of measurement, gravimetry improves our capability of analyzing the factors that cause variations in sea level as well as the regional effects those factors have.

3. **The Consequences for the Evaluation of Climate Change**

The Process of Monitoring and Projecting

It is essential to have measurements of the rise in sea level that are both accurate and trustworthy in order to keep track of the changes that are now taking place and to forecast what the future may hold. Researchers are able to construct more accurate models to estimate the possible effects of rising sea levels on coastal communities, infrastructure, and ecosystems if they combine the data obtained from a number of different measurement techniques. These projections are essential for understanding adaptation and mitigation strategies for climate change at the local, regional, and global scales.

Formulation of Policies and Procedures

Measurements of the rise in sea level are extremely important in the formation of policies about climate change and the informing of decision-making processes on both the national and international levels. Accurate and up-to-date data on sea level rise are essential to the success of policy programs that aim to cut emissions of greenhouse gases, safeguard coastal areas through the implementation of various protection measures, and support sustainable development. In order to effectively advocate for policies that address the myriad of issues that are being posed by rising sea levels and the attendant repercussions, solid scientific data is absolutely necessary.

The Socioeconomic Consequences

For the purpose of determining the social and economic effects of climate change, it is essential to have a solid grasp of sea level rise observations. Sea level rise poses substantial threats to vulnerable coastal towns, particularly those located in low-lying and developing regions. These threats include the dislocation of residents, the destruction of infrastructure, and economic losses. It is crucial to have access to accurate data regarding sea level in order to establish measures that will increase community resilience, promote equitable adaptation, and ensure the socioeconomic well-being of communities that are impacted.

Cooperation across National Boundaries

The observations of the rise in sea level and the implications of these measurements highlight the necessity of worldwide collaboration and cooperation in order to meet the global problem posed by climate change. Initiatives such as the Intergovernmental Panel on Climate Change (IPCC) serve as venues for the exchange of scientific knowledge, the promotion of data transparency, and the facilitation of collaborative efforts to develop efficient measures for mitigating the effects of climate change. Increased international cooperation is absolutely necessary in order to handle the multifaceted and interrelated problems that are associated with rising sea levels and the broader implications that this phenomenon has for the global society.

The rise in sea level is an important part of climate change, and in order to effectively assess and respond to it, one has to have a full understanding of it as well as precise observations. When many methods of measurement, such as tide gauges, satellite altimetry, and gravimetry, are combined, a full view of changes in sea level as well as the regional differences that accompany those changes is provided. These measurements are of critical importance for informing assessments of climate change, directing the development of policies, and supporting decision-making processes on local, national, and international levels. Building a more resilient and sustainable future for all parties involved requires a concerted effort to improve sea level measuring skills and strengthen interdisciplinary collaboration. This is especially important as the international community continues to struggle with the problems posed by climate change.

6.2 The link between melting ice and rising seas

One of the most obvious and undeniable effects of global warming is the correlation between the retreat of ice sheets and an increase in water levels in coastal areas. Because of an increase in average temperatures around the world, the ice sheets, glaciers, and polar caps on Earth have been melting at an accelerated rate over the course of the past century. This phenomena has significant repercussions for the ecosystems of the world's oceans, as well as for coastal towns and the overall climate system of the entire globe. This article presents a comprehensive examination of the connection between melting ice and increasing sea levels. It investigates the causes, effects, and potential mitigation techniques for this significant element of climate change.

1. Factors That Contribute to Ice Melting

Warming of the Planet

The primary factor that contributes to the melting of ice is global warming, which is caused by a rise in the amount of greenhouse gases that are present in the atmosphere of the Earth. Carbon dioxide (CO2) and other greenhouse gases are released into the atmosphere when human activities such as the burning of fossil fuels, the clearing of forests, and industrial operations are carried out. These gases have a heat-trapping effect, which contributes to a general warming of the earth. As average temperatures around the world continue to rise, the effects that this has on ice masses will become more obvious.

Warming of the Oceans

Another key element that contributes to the melting of ice is the warming of the ocean's surface temperatures. This is especially true for glaciers and ice shelves that are in direct contact with the water. Warm ocean waters have the ability to damage the undersides of ice shelves, which can lead to the shelves becoming unstable and possibly collapsing. The melting of glaciers is accelerated by increased oceanic heat, which in turn causes freshwater to be released into the ocean, which in turn causes the sea level to rise.

The use of Feedback Loops

The process of ice melting itself can trigger feedback loops that make the ice loss problem more worse. For instance, as ice sheets melt, they can reveal dark surfaces such as bare land or water, both of which absorb more heat from the sun than ice and snow can. This can cause temperatures to rise. In turn, this leads to a faster overall warming and an increase in the amount of melting. In addition, as glaciers recede, they might uncover previously buried ice layers, which can lead to additional melting and the release of freshwater into the ocean. This process occurs as the glaciers retreat.

2. **Different Ways That Ice Can Melt**

The Ice Caps at the Poles

The ice caps in the polar regions, such as the Arctic and the Antarctic, are currently seeing some of the most severe melting. The Arctic in particular has been warming at a rate that is more than double the norm for the rest of the world, which has led to a significant drop in the amount of sea ice. The melting of Arctic sea ice has a domino impact on the entire climate system, including changes in the patterns of ocean circulation, which can have an effect on the climate of other regions as well as the entire planet.

The West Antarctic Ice Sheet (also known as the WAIS) and the Antarctic Peninsula are two places of the Antarctic that are of special concern. In particular, the WAIS has revealed indications of major ice loss, which, if accelerated, might contribute significantly to an increase in the level of seawater around the world.

Glaciers

Another significant cause of ice loss is the melting of mountain glaciers, which can be found in many different parts of the world. Glaciers are an extremely useful indication of climate change due to the fact that their size and extent are closely tied to the weather conditions in their respective regions. The alarming rate at which mountain glaciers are receding as a result of rising temperatures is contributing to the flow of meltwater into rivers and, in turn, to an increase in the level of seawater.

The Ice Sheet of Greenland

There is a significant quantity of freshwater that is frozen inside of the Greenland Ice Sheet, which is the world's second-largest ice body and the world's second-largest ice sheet overall. The Greenland Ice Sheet has been experiencing a rate of surface melting that is accelerating over the past few decades, with surface melt events growing more frequent and broad. Because the meltwater coming from Greenland contributes directly to the rise in sea level, it has become the focus of intensive inquiry and concern among scientists.

3. **The Repercussions of the Ice Melting**
Increasingly High Sea Levels
The rise in sea levels is the most immediate and direct impact that can be attributed to the melting of glaciers. Freshwater is added to the oceans around the world as a result of the melting of ice that comes from glaciers, polar regions, and ice sheets. Since the late 19th century, the average level of the sea throughout the world has increased by around 20 cm (about 8 inches), and the rate at which the sea level is rising is accelerating. According to some projections, global sea levels might rise by as much as 1 meter (3.3 feet) by the end of the 21st century if we continue on our current path of producing a lot of greenhouse gases.

Erosion of the Coastline
Coastal erosion is exacerbated by rising sea levels because higher water levels make shorelines more susceptible to both more frequent and more intense wave activity. Erosion places coastal towns in a more precarious position, making them more likely to suffer the loss of land, property, and infrastructure. In certain areas, entire coastal towns and cities are in danger of being destroyed or rendered uninhabitable because of rising sea levels.

A rise in the level of flooding
The rise in sea level makes coastal flooding more severe when it occurs in conjunction with storms and high tides. Even very modest rises in sea level can considerably amplify the destructive

potential of storm surges, which in turn can cause more frequent and severe flooding along coasts. Major coastal cities such as Miami, New York City, and Shanghai are particularly vulnerable to the effects of climate change.

The Uprooting of Whole Communities

As a result of rising sea levels, coastal populations, particularly those located in low-lying areas, are at risk of being displaced. This is especially worrisome for geographically compact island nations and indigenous groups that are highly dependent on coastal areas for their survival. It is possible for people to lose their homes, their livelihoods, and their cultural heritage as a result of displacement, which creates severe social and economic issues.

Infiltration of Saltwater

Salt water can seep into freshwater sources, particularly coastal aquifers, if sea levels continue to rise at their current rate. Because of this, agricultural production and the supply of clean drinking water are both put in jeopardy, as land rendered infertile by salt-contaminated groundwater also impacts the availability of clean drinking water.

Influence on the World's Ecosystems

Coastal ecosystems, such as mangroves, estuaries, and wetlands, are extremely vulnerable to the rise in sea level. These habitats are at risk of becoming submerged as a result of rising sea levels, which will result in the loss of habitat and will have an impact on the species that rely on these ecosystems. The destruction of coastal ecosystems can have a negative impact on fisheries, the quality of the water, and the resistance of coastal areas to the effects of natural disasters.

4. Strategies for Adaptation and Mitigation of Impacts

Combining methods for climate change adaptation and mitigation is necessary in order to address the implications of ice melting and rising sea levels.

Adjustment (Mitigation)

1. Reducing carbon emissions by shifting to more environmentally friendly and renewable forms of energy.
2. Improving energy efficiency with the goal of lowering total energy usage.
3. The planting of new trees and the maintenance of existing forests in order to sequester and store carbon.
4. Technologies known as carbon capture and storage (CCS), which may remove carbon dioxide from the atmosphere.

The process of adapting

The primary goals of adaptation methods are to mitigate the effects of rising sea levels and to make necessary preparations for them.

1. Constructing new infrastructure or modifying existing structures to better withstand the effects of rising flooding and erosion.
2. The establishment of early warning systems in order to provide communities with information regarding imminent coastal risks.
3. Zoning rules and land-use planning to limit growth in coastal areas that are susceptible to erosion.
4. Practices of sustainable land use, such as the maintenance of green spaces and buffer zones, which are able to absorb excess water during storms.

Cooperative efforts on a global scale

In order to effectively handle the worldwide problem of glaciers melting and sea levels rising, international cooperation is very necessary. Platforms such as the United Nations Framework Convention on Climate Change (UNFCCC) and the Intergovernmental Panel on Climate Change (IPCC) give opportunities for countries to collaborate on the reduction of global warming and the development of plans for adaptation. Increased international cooperation is absolutely necessary in order to solve the multifaceted and interrelated problems that are involved with the melting of ice and the broader consequences that this phenomenon has for the global society.

Investigation and Observation

Continued research and monitoring of ice melt and its repercussions are vital for understanding the shifting issues and determining how well adaptive and mitigation efforts are working. This knowledge can help shape public policy, provide support for decision-making, and raise awareness among the general public.

One of the most important aspects of the current climate crisis is the correlation between ice melting and rising sea levels. Ice from polar regions, glaciers, and ice sheets is melting at an accelerated rate as global temperatures continue to increase. This is leading to the flooding of coastal areas all over the world. The importance of tackling this world-wide problem is highlighted by the consequences of rising sea levels, which include coastal erosion, flooding, the relocation of towns, the incursion of saltwater, and changes to ecosystems.

Building resiliency in the face of climate change requires implementing both adaptation and mitigation methods. Mitigation efforts aim to lower emissions of greenhouse gases, while adaptation strategies prepare for and lessen the negative effects of rising sea levels. Cooperation on a worldwide scale is essential if we are to handle the multifaceted and interrelated problems that are connected to the melting of ice and the broader consequences that this phenomenon has for the global society.

As the international community continues to struggle with the effects of climate change, it is absolutely necessary to make a determined effort to strengthen interdisciplinary collaboration and better knowledge of the connection between melting ice and rising seas in order to construct a future that is more robust and sustainable for everyone.

6.3 Coastal regions and cities most vulnerable to sea level rise

The rise in sea level, which is being caused by global warming and the melting of polar ice, glaciers, and ice sheets, is a significant risk to coastal areas and towns all over the world. It is possible that rising sea levels will cause low-lying areas to become submerged, will damage coastlines, and will affect the lives and businesses of millions of people. In this article, we will investigate the coastal cities and regions that are most at risk from the rising sea level. We will investigate the primary causes that contribute to their susceptibility, as well as the potential repercussions and the efforts that are now being made to adapt to and minimize the effects of this global challenge.

1. **Extremely Vulnerable Coastal Areas**
   Coastal Regions That Have a Low Elevation
   The rise in sea level poses the greatest threat to low-lying coastal areas, which are frequently found at or very close to sea level. Plains that are formed by deltas, estuaries formed by rivers, and coastal wetland areas are included in these regions. Some notable examples of low-lying coastal areas include the Ganges-Brahmaputra Delta in India and Bangladesh, the Nile Delta in Egypt, and the Mississippi Delta in the United States.
   Small countries and island groups
   Small island nations are particularly vulnerable to the effects of sea level rise since their highest points are sometimes only slightly elevated above the water. Typical instances include nations located in the Pacific Ocean, such as Tuvalu, Kiribati, and the Marshall Islands. These countries are in danger of being submerged by rising sea levels, which would render them increasingly uninhabitable.
   Asia in the Southeast
   The low-lying coastal regions of Southeast Asia are home to a sizeable population that call such regions their home. Countries such as Vietnam, Cambodia, and Thailand are especially susceptible to the effects of climate change. The Mekong Delta, for example, is at a high risk because of its wide network of rivers and marshlands, which makes it vulnerable to both an increase in the level of the sea and changes in the flow of the river farther

upstream.

Coastal Areas that Have a High Population Density

Coastal areas that have a high population density are at a greater risk than other coastal areas because there is a higher concentration of people, infrastructure, and economic activity in these areas. Cities such as Mumbai, Jakarta, and Dhaka are prime examples of the potential effects that rising sea levels could have on highly populated locations. In these cities, millions of citizens run the risk of being displaced and their economies could suffer as a direct result.

2. **Cities in Dangerous Coastal Areas**

The United States of America, Miami (Florida)

One of the most iconic cities in the world that is under danger because to rising sea levels is Miami, which is located in South Florida. Because of its low-lying topography and porous limestone bedrock, the city is especially susceptible to natural disasters. The rise in sea level, when combined with the storm surges that accompany storms, represents a significant risk to Miami's infrastructure, especially the city's crucial tourism sector.

America's New York City

Because of its wide shoreline and low-lying coastal districts, New York City is extremely vulnerable to both rising sea levels and harsh weather occurrences. The devastation caused by Superstorm Sandy in 2012 served as a jarring illustration of how vulnerable the city is. The extensive floods and damage to infrastructure that occurred as a result of that storm highlight the necessity of developing thorough adaption methods.

China: Shanghai (Shanghai)

Sea level rise poses a huge threat to Shanghai, which is not only one of the largest cities in China but also an important economic hub around the world. The low-lying coastal sections of the city are home to a sizeable majority of the city's population. The infrastructure of the city, as well as the industrial districts and port facilities, are all under risk from the rising sea levels.

The Thai city of Bangkok

Due to the extraction of groundwater, the city of Bangkok, which is situated on the Chao Phraya Delta, is threatened not only by the rising sea level but also by subsidence. As was seen during the floods that occurred in 2011, the city has a significant risk of being flooded. The government of Thailand has begun work on a number of initiatives to address the issues posed by rising sea levels.

3. Contributing Factors That Make People Vulnerable

Having a low elevation

A significant factor that contributes to coastal cities and regions' susceptibility to the effects of rising sea levels is the elevation of those areas. The most vulnerable locations are low-lying areas that have elevations that are either close to or below sea level. Even moderate rises in sea levels have the potential to cause widespread flooding and inundation of coastal areas in these areas.

Density of the Population

Because there is a higher concentration of people, infrastructure, and economic activities in coastal cities with high population densities, these cities are more susceptible to the negative effects of rising sea levels. The displacement of people and the loss of economic opportunities in highly populated metropolitan regions can have far-reaching effects on both society and the economy.

The study of geology

The geological make-up of coastal places can have an effect on the degree to which they are vulnerable. When subjected to a rise in sea level and an increase in the intensity of storm surges, regions that have porous geological formations, such as limestone or sandy soils, may experience more rapid subsidence and coastal erosion.

Closeness to various Waterways

Cities and regions that are located near beaches, river estuaries, or in close proximity to waterways are at a greater risk of being negatively affected by rising sea levels and the repercussions that come along with it. These places are vulnerable to storm surges, increasing tidal floods, and the intrusion of saltwater into freshwater sources.

4. **Possible Repercussions of This Action**

Erosion of the Coastline

Coastal erosion is one of the most immediate repercussions that will arise as a result of rising sea levels. Increasing sea levels make wave activity along shorelines more frequent and violent, which results in the erosion of land and the destruction of infrastructure. Because it can damage beaches and other natural attractions, coastal erosion is a major source of concern for towns and regions that rely heavily on tourism as an economic driver.

A rise in the level of flooding

The rise in sea level is a contributing factor in the worsening of coastal flooding caused by storms and high tides. Even modest rises in sea level can have a significant impact on the severity of storm surges, which in turn can cause more frequent and destructive flooding along coasts. Floods have the potential to obstruct transportation, cause damage to property, and put citizens' health and safety at risk.

The Uprooting of Whole Communities

Coastal towns, particularly those located in low-lying areas, may be required to relocate as a consequence of rising sea levels.

When people are displaced, they lose their homes, their livelihoods, and their cultural heritage. It may be necessary in some instances to uproot entire communities and move them to more secure locations; this might provide a number of difficult social and economic issues.

Infiltration of Saltwater

Salt water can seep into freshwater sources, particularly coastal aquifers, if sea levels continue to rise at their current rate. This can lead to the infertility of land, the disruption of agricultural practices, and an impact on the supply of clean drinking water. It poses a considerable risk to the agricultural production as well as the water supplies used for drinking in coastal areas.

Influence on the Economy

The rise in sea level can have significant repercussions for economies around the world. Ports, industrial facilities, and other essential infrastructure can frequently be found in coastal regions. It is possible for these assets to cause considerable economic losses if they are disrupted. Damage to ports, for instance, can impede commerce and supply chains, which in turn can have an impact on both local and global economies.

Influence on the World's Ecosystems

Coastal habitats, such as mangroves, estuaries, and wetlands, are extremely susceptible to fluctuations in sea level. The loss of habitat can have a negative impact on the populations of animals and plants that are dependent on particular ecosystems. The destruction of coastal ecosystems can have a negative impact on fisheries, the quality of the water, and the resistance of coastal areas to the effects of natural disasters.

5. **Efforts to Adapt to and Mitigate the Impacts of the**

Coastal Defense and Defense Measures

The effects of coastal erosion and floods can be mitigated to some degree through the implementation of coastal protection measures such as seawalls, levees, and the restoration of sand dunes. These constructed solutions give temporary relief, but they should be included as part of a larger strategy that also includes natural solutions such as the restoration of mangrove forests.

Planning for the Sustainable Use of Land and Urban Space

Coastal cities and regions can be made less susceptible to natural hazards through the implementation of sustainable land management methods and urban planning. This involves the preservation of green spaces, wetlands, and buffer zones that have the capacity to absorb surplus water during periods of precipitation.

It is also essential to construct physical infrastructure that takes into account future rises in sea level, such as roads and buildings that are raised.

Planning for Adaptations

In order to address both the short-term and the long-term effects of increasing sea levels, coastal towns and governments need to engage in extensive adaptation planning. This includes taking precautions such as constructing or modifying infrastructure so that it can resist floods, implementing early warning systems, and enacting zoning restrictions that restrict development in regions that are susceptible to flooding.

Regression and Repositioning

Communities that are at a high risk due to rising sea levels may find that retreat and relocation are the most practical options for them in some circumstances. This necessitates the relocation of residents away from dangerous coastal areas and into more secure areas. To ensure the well-being of communities that have been uprooted, rigorous planning and support are both necessary.

Cooperative efforts on a global scale

In order to effectively address the worldwide problem of rising sea levels, international cooperation is very necessary. Platforms such as the United Nations Framework Convention on Climate Change (UN-FCCC) and the Intergovernmental Panel on Climate Change (IPCC) give opportunities for countries to collaborate on mitigating the effects of global warming and devising plans to adjust to those effects.

Investigation and Observation

Understanding the developing difficulties and determining how well adaption methods are working requires ongoing research and monitoring of the rise in sea level and its repercussions. This knowledge can help shape public policy, provide support for decision-making, and raise awareness among the general public.

Coastal areas and cities that are most susceptible to the effects of sea level rise are confronted with a variety of difficult problems that are intricately tied to one another. These problems include coastal erosion, increasing floods, the relocation of communities, saltwater intrusion, and economic losses. The effects of rising sea levels highlight the critical need to address this global crisis as quickly as possible and put into action a mix of measures that address both adaptation and mitigation.

The complex and intertwined problems that are related with rising sea levels and the broader implications that this phenomenon has for the global society can only be effectively addressed via the participation of the international community. For the sake of creating a more resilient and sustainable future for everyone, it is absolutely necessary to make a concentrated effort to strengthen inter-disciplinary collaboration and gain a better knowledge of the vulnerabilities of coastal regions and cities.

Chapter 7

Ocean Acidification

Ocean acidification is a significant environmental issue that has far-reaching ramifications for marine ecosystems and global biodiversity as a result of increased carbon dioxide (CO2) absorption by the world's oceans. This is a consequence of increased carbon dioxide (CO2) absorption by the world's oceans. The seas absorb a substantial percentage of this surplus carbon dioxide as atmospheric CO2 levels continue to rise due to human activities such as the burning of fossil fuels and deforestation; as a result, changes in the chemistry of saltwater and a reduction in pH are occurring as a result of these absorptions. This article offers a full examination of ocean acidification, delving into its causes as well as the ecological and economic repercussions that have resulted from it. It also discusses prospective techniques for mitigating the effects of ocean acidification, which is a crucial part of global environmental change.

1. **An Awareness of the Acidification of the Ocean Procedures involving Chemistry**
   The primary contributor to ocean acidification is the dissolution

of carbon dioxide (CO2) in saltwater, which results in the production of carbonic acid. As a result of this process, the concentration of hydrogen ions in the water will grow, which will cause the pH to decrease. Because many marine creatures, such as corals, mollusks, and certain forms of plankton, depend on a constant pH environment for their survival and the calcification processes, the chemical reactions that are involved in ocean acidification can have substantial repercussions for marine life.

The pH Scale and Its Various Levels of Acidity

The pH scale has a range from 0 to 14, and a value of 7 is regarded to be neutral. This scale is used to assess how acidic or alkaline a solution is. The pH values of seawater are decreasing, which indicates a trend toward increased acidity as ocean acidification continues to occur. Even though the pH of the world's oceans is falling, they are still regarded to be alkaline. However, this poses a considerable risk to marine ecosystems, particularly those that are dependent on carbonate ions for the development of shells and skeletons.

The Worldwide Carbon Cycle

The earth's oceans act as a huge carbon sink, taking up excess carbon dioxide from the air and storing it for future use. This is an essential part of the global carbon cycle. However, the rising levels of carbon dioxide in the atmosphere have caused a disruption in the natural balance of carbon, which has led to the acidification of seawater and has had a severe influence on marine life. It is vital to have a solid understanding of the linked nature of the global carbon cycle in order to have a complete comprehension of the far-reaching effects that ocean acidification has on marine ecosystems.

2. **The Factors That Contribute to Ocean Acidification Increasing CO2 Levels in the Atmosphere**

The principal factor contributing to the acidification of the ocean is the rising concentration of CO2 in the atmosphere, which is caused by human activity. Large volumes of carbon dioxide (CO2) are released into the atmosphere whenever fossil fuels are burned, forests are cut down, or industrial operations are carried out. As a large portion of this excess CO2 is absorbed by the seas, the increasing concentration of carbonic acid results in a reduction in the pH of the seawater, which poses a threat to marine life, particularly those marine organisms that rely on calcium carbonate for their skeletal systems.

Carbon Cycles Found in Nature

Although human activities are a major contributor to ocean acidification, natural carbon cycles, such as volcanic activity and the breakdown of organic matter, also play a part in the release of CO2 into the atmosphere. Human activities are the primary cause of ocean acidification. However, the current rate of CO2 emissions that are the consequence of human activities has far surpassed the capacity of natural carbon sinks to absorb carbon dioxide and buffer the effects of ocean acidification. This is because human activities produce a much higher rate of CO2 emissions.

The use of Feedback Loops

The acidity of the ocean has the potential to set off feedback loops that will further intensify the damage it does to marine eco-systems. The inability of some marine animals to produce shells and skeletons as a result of decreased pH levels, for instance, can disrupt food chains and the dynamics of ecosystems, leading to repercussions that cascade throughout the marine food web.

3. **The Effects That Ocean Acidification Has On The Environment**

Influence on the Organisms of the Sea

The acidification of the ocean presents a severe risk to marine life, particularly to creatures that

construct their shells and skeletons out of calcium carbonate and are therefore dependent on the ocean's pH. Because of the reduction in seawater pH, it is more difficult for organisms such as corals, mollusks, echinoderms, and certain kinds of plankton to create and maintain their calcified structures. As a result, these organisms are especially susceptible to the consequences of ocean acidification. The weakening of these structures can render the creatures more vulnerable to being preyed upon as well as being subjected to the stresses of their environment.

Causes of Disturbance to Marine Ecosystems

The upheaval of marine ecosystems that might arise from the acidification of oceans can have ecological repercussions that are far-reaching. Because of the decrease in pH levels, coral reefs, sometimes known as the "rainforests of the sea," are especially susceptible to the negative impacts of ocean acidification. This is because the decrease in pH levels prevents coral from calcifying and growing. This can result in the destruction of essential habitat for a wide variety of marine species, having a negative influence on both the biodiversity and the stability of the ecosystem.

Implications for the Economy

The acidity of the ocean could have substantial repercussions for the economy, in particular for businesses that are dependent on the integrity of marine ecosystems, such as those involved in fishing, aquaculture, and tourism. The decrease in shellfish populations, for instance, can cause disruptions in the seafood sector, which can result in economic losses and the potential loss of livelihoods for coastal communities that are dependent on fishing and shellfish harvesting.

Influences on the Availability of Food

The dislocation of marine food webs that can be attributed to ocean acidification can have significant repercussions for the safety of the world's food supply. Particularly for communities that rely on marine resources as their primary source of nourishment, the decrease in population sizes of essential marine species, like as fish and shellfish, can have an effect on the availability of seafood that is high in protein.

4. **Strategies for Adaptation and Mitigation of Impacts**

Strategies for Risk Minimization

1. **Making the Switch to Clean Energy:** Moving away from fossil fuels and encouraging the use of renewable energy sources is one approach to dramatically cut CO2 emissions and slow down the rate at which ocean acidification occurs.

2. **Carbon Capture and Storage (CCS):** The implementation of CCS technology can capture CO2 emissions from industrial processes and power plants, preventing them from entering the atmosphere and, as a result, being absorbed by the seas. These emissions can be stored in underground reservoirs.

3. **Reforestation and Afforestation:** Efforts made toward reforestation can assist absorb CO2 from the atmosphere, so reducing the overall carbon footprint and mitigating the effects of ocean acidification. Afforestation is the practice of planting trees in areas that have been deforested.

Adaptation Methods and Techniques

1. **Ecosystem-Based Adaptation:** The implementation of ecosystem-based adaptation measures, such as the restoration of coastal habitats and the development of marine protected areas, can help strengthen the resilience of marine ecosystems to the effects of ocean acidification. These strategies include the creation of marine protected areas.

2. **methods of Sustainable Aquaculture:** The promotion of methods of sustainable aquaculture that minimize the environmental impacts of fish and shellfish farming can assist in lowering the

overall susceptibility of marine creatures to the effects of ocean acidification.

3. **study and Monitoring:** Continuing study and monitoring of ocean acidification and its repercussions are vital for understanding the growing problems and determining effective solutions for adaptation.

Cooperative efforts on a global scale

1. International Cooperation and the Exchange of Knowledge The need for international cooperation and the exchange of knowledge is vital in order to solve the worldwide problem of ocean acidification. Initiatives such as the Paris Agreement, which aims to limit global warming, provide a forum for governments to work together in minimizing the impacts of ocean acidification and supporting sustainable ocean management practices. The Paris Agreement's primary objective is to limit the temperature increase caused by human activities.

The acidification of the ocean is a huge environmental problem that poses considerable risks to marine ecosystems, global biodiversity, and the way of life of populations that are reliant on marine resources for their livelihoods. The increase in atmospheric CO2 levels that is a direct result of human activity has hastened the acidity of the world's oceans, which has led to far-reaching effects for both the environment and the economy. It will take a coordinated effort to reduce emissions of carbon dioxide (CO2), to promote sustainable ocean management practices, and to strengthen the resilience of marine ecosystems in order to address the difficulties posed by ocean acidification. Cooperation on an international level and ongoing research are two things that are very necessary for the development of efficient strategies for mitigating and adapting to climate change in order to guarantee the long-term health and sustainability of the world's oceans and the communities that are dependent on them.

7.1 Explanation of the ocean's absorption of excess CO2

The oceans, which make up more than 70 percent of the surface of the Earth, are extremely important participants in the process of controlling the global carbon cycle. One of the many services that they perform, one of the most important of which is the removal of excess carbon dioxide (CO2) from the air, is carried out by trees. Carbon sequestration is a method that helps minimize the effects of climate change by lowering the quantity of carbon dioxide (CO2) in the atmosphere. CO2 is a primary greenhouse gas that is responsible for global warming. In this post, we will go into the explanation of how the ocean is able to absorb extra carbon dioxide, the elements that influence this process, and the importance of understanding and regulating this critical ecosystem service.

1. The Part Played by the Ocean in the Locking Up of Carbon

An Overview of the Carbon Cycle

1. **Carbon Reservoirs:** These are regions that can store carbon, such as the atmosphere, seas, terrestrial vegetation, and soil. b. Carbon Sinks: These are areas that can release carbon into the atmosphere.
2. **Carbon Fluxes:** The processes that move carbon between different reservoirs are referred to as carbon fluxes. The processes of photosynthesis and respiration are two examples of carbon fluxes. Another example is the transfer of carbon dioxide from the atmosphere to the oceans.

The Role of the Ocean as a Carbon Sink

1. **Solubility Pump:** Because of the disparity in CO_2 concentrations between the atmosphere and the surface waters of the ocean, the ocean's surface waters quickly absorb CO_2 from the atmosphere. This process is affected by temperature and pressure, with fluids that are colder and deeper having the capacity to hold more dissolved carbon dioxide.

2. **Biological Pump:** Phytoplankton and other marine species, in addition to playing an important

part in the carbon cycle, are also known as the biological pump. These creatures fix atmospheric CO_2 into organic matter through the process of photosynthesis. This organic matter can sink to the depths of the ocean when these species die or when other organisms consume them.

II. Factors That Play A Role In The Absorption Of Carbon By The Ocean

Degrees Fahrenheit

Temperature plays a role in whether or not CO_2 may be dissolved in seawater. When compared to warmer water, colder water has a greater capacity to store dissolved carbon dioxide. As a consequence of this, polar regions and the waters found deep within the ocean are better able to store CO_2 due to the lower temperatures found there.

Mixing at the Surface

There is a potential for increased CO_2 transfer between the upper ocean layers and the atmosphere through the facilitation of physical processes such as wind-driven surface mixing and upwelling. Because of this mixing, the ocean's potential to take in carbon dioxide from the atmosphere is increased.

Biological Actions or Processes

Marine organisms, phytoplankton in particular, play a significant role in the process of carbon sequestration that is carried out by the biological pump. The amount of available nutrients, the amount of light, and the temperature all play a role in the growth of phytoplankton.

Increased nutrient inputs have the potential to encourage the growth of phytoplankton, which will, in turn, result in increased carbon sequestration.

The flow of blood

The migration of CO2-rich surface waters down into the deeper ocean layers is influenced by the circulation patterns of the ocean. The movement of dissolved carbon dioxide to deeper oceanic depths is one step in a longer-term process called carbon sequestration, which stores carbon for protracted periods of time.

III. Sequestration of Carbon in the Depthes of the Oceans

Mixing in a Vertical Array

The process of vertical mixing of ocean waters involves the transfer of surface waters, which are rich in carbon dioxide, into deeper layers of the ocean. The surface temperature, the winds, and the thermohaline circulation are the primary forces that drive this process. As surface waters descend, they carry dissolved carbon dioxide to greater depths, which helps to contribute to the long-term storage of carbon.

That Which Is Organic

The biological pump, which is driven by marine creatures, is one factor that contributes to the deep ocean's ability to sequester carbon. When phytoplankton and other creatures pass away or are consumed, the organic matter that they produce can sink to the seafloor. This has the effect of taking carbon from the surface of the ocean and depositing it in the deeper waters below.

Carbon Sequestration

Carbon that is stored in the abyssal and hadal zones of the ocean is said to be able to remain there for thousands of years. Other deep ocean layers also function as stores for carbon. In these areas, the long-term retention of carbon is helped along by a combination of factors including pressure, low temperature, and decreased microbial activity.

IV. The Importance of Comprehending the Carbon Sequestration Process in the Ocean

Regulation of the Climate

In order to effectively manage climate change, one must have a comprehensive understanding of how carbon is stored in the ocean. Because they take in and store excess carbon dioxide, the seas contribute to the mitigation of the effects of global warming. For the purposes of accurately projecting future climatic scenarios and guiding climate policy, accurate observations and modeling of this process are key components.

Wellness of Ecosystems

The ability of marine ecosystems to sequester carbon is directly correlated to how healthy and resilient they are. It is absolutely necessary to have a solid understanding of the components that determine CO2 uptake, such as temperature and biological activity, in order to properly evaluate and manage the effects of climate change on marine life.

Carbon Expenditure Plan

It is possible to arrive at a more precise estimate of the planet's carbon budget if one has a thorough grasp of the process of carbon sequestration in the ocean. Calculations of the carbon budget that are accurate provide policymakers with the information they need to establish successful measures to restrict CO2 emissions and lessen the overall effects of climate change.

Acidification of the Ocean

Ocean acidification is a consequence of the ocean's role in carbon sequestration, which, although necessary for maintaining a stable climate, also contributes to ocean acidification. For the purpose of formulating solutions to this significant environmental problem, it is essential to have a solid understanding of the dynamic relationship that exists between carbon uptake and ocean acidification.

V. Obstacles and Causes for Concern

Acidification of the Ocean

The ocean becomes more acidic as a result of the absorption of carbon dioxide by seawater, which can have negative consequences for marine life. The inability of organisms to produce shells and skeletons, as well as the disruption of food chains and the threat to biodiversity, are all caused by increased acidity. The problem of reducing the negative effects of ocean acidification is one that is both complex and urgent.

Oceans Getting Warmer

The ability of the ocean to store carbon could be impaired if temperatures continue to rise. The solubility of carbon dioxide could decrease as water temperatures rise, which could eventually result in a reduction in the amount of carbon that is absorbed. Additionally, warmer temperatures have the potential to change marine habitats, which in turn can have an effect on how effectively the biological pump works.

A state of eutrophication

Eutrophication is a process that can occur in coastal areas as a result of an excessive inflow of

nutrients, most frequently from agricultural runoff and pollution. In the short term, this can boost the growth of phytoplankton and lead to carbon sequestration, but over the longer term, it can lead to oxygen-depleted "dead zones" and blooms of toxic algae species.

Alterations in the Ocean Circulation

Alterations in the patterns of ocean circulation, which may be brought on by climate change, may have an impact on the distribution of dissolved carbon dioxide.

Changes in the patterns of circulation in the ocean have the potential to affect the movement of carbon-rich surface waters to deeper ocean layers.

VI. Preventative Measures and Emergency Planning
Bringing Down Our CO2 Levels

Reducing emissions of carbon dioxide at the point where they are produced is the most efficient strategy to address the problems associated with ocean carbon sequestration. Essential initiatives include making the switch to clean and renewable energy sources, improving energy efficiency, and decreasing dependency on fossil fuels.

Management of the Ocean That Is Sustainable

It is absolutely necessary to put into practice methods of ocean management that are sustainable if one want to preserve the vitality and resistance of marine ecosystems. The construction of marine protected areas, the reduction of nutrient pollution, and the regulation of fishing and aquaculture practices are all strategies that have been implemented.

The acronym "Capture and Storage of Carbon" (CCS)

The research, development, and deployment of carbon capture and storage (CCS) technologies can assist in the capture and storage of CO2 emissions from industrial processes and power plants, thereby preventing those emissions from being released into the atmosphere and then being absorbed by the oceans.

Investigation and Observation

It is necessary to maintain research on, and monitoring of, ocean carbon sequestration and its implications in order to gain a better knowledge of the emerging difficulties and to guide appropriate methods for mitigation and management. Research in the scientific realm is absolutely necessary to both the broadening of our knowledge base and the production of novel solutions.

The process by which excess carbon dioxide is absorbed by the ocean is an essential step in the carbon cycle of the Earth. This step has significant repercussions for the control of the climate as well as the integrity of marine ecosystems. Understanding the processes that influence carbon sequestration in the oceans is vital for regulating the effects of climate change, mitigating ocean acidification, and preserving the long-term sustainability of our planet as CO2 emissions from human activities continue to grow.

The implementation of measures to lower CO2 emissions, the promotion of sustainable ocean management, and investments in technology that collect and store carbon are critical actions that must be taken in order to preserve the priceless contribution that the oceans provide toward the mitigation of climate change.

7.2 Impacts on marine life, including coral reefs and shellfish

Ocean acidification is a serious environmental issue that has far-reaching ramifications for marine ecosystems. It is a consequence of rising atmospheric carbon dioxide (CO2) levels and increased CO2 absorption by the world's oceans. Oceans absorb more CO2 than ever before. In this article, we will discuss the effects that ocean acidification has on marine life, with a particular emphasis on two groups that are particularly susceptible to its effects: coral reefs and shellfish. Because of the heightened sensitivity of these species and many others like them to shifts in the chemical make-up of the ocean, we can use them as markers of the far-reaching implications of this worldwide problem.

1. **The Acidification of the Ocean and Its Effects on the Chemistry of the Sea**

   The Dissolution of Carbon Dioxide

   The fundamental factor that contributes to the acidity of the ocean is the solubility of carbon dioxide in seawater. Carbonic acid is produced when CO2 from the atmosphere is absorbed by the oceans because this causes it to react with the water present. As a result of this reaction, the concentration of hydrogen ions (H+) in the water increases, which in turn causes the pH of the environment to drop, making it more acidic.

   The pH Scale and the Acidification of the Ocean

   The degree of acidity or alkalinity of a solution can be determined using the pH scale. The pH scale is scaled from 0 to 14, with 7 representing neutrality, 0 representing acidity, and 7+ representing alkalinity. The pH of seawater will drop as a result of ocean acidification, which will bring it closer to the more acidic end of the scale. The consequence of this fall in pH is important, particularly for marine creatures that depend on stable pH levels for a variety of physiological activities. These organisms are particularly vulnerable to the effects of this decrease in pH.

2. **Impacts on the World's Coral Reefs**

   The Study of Corals

   Coral reefs, also known as the "rainforests of the sea," are ecosystems that are notable for their high levels of biodiversity and ecological significance.

Coral polyps, which release calcium carbonate to form their skeletons, are responsible for their construction. These reefs offer a safe haven, food, and a habitat to a wide variety of marine life, in addition to contributing to the protection of the coastline.

Exposure to the Hazards of Ocean Acidification

As a result of their dependency on calcium carbonate for the development of their skeletons, coral reefs are extremely susceptible to the negative impacts of ocean acidification. Coral calcification is hampered when pH levels continue to fall, which is detrimental to the structural integrity of reefs. This precarious situation is made even more precarious by the fact that pollution and rising ocean temperatures are both contributing factors.

Events That Cause Bleaching

The phenomena of coral bleaching, which is caused by the warming of the water, is made worse by the acidification of the ocean. When coral polyps expel their zooxanthellae, often known as their symbiotic algae, they not only lose their color but also become more susceptible to disease and eventually die. Coral reefs are experiencing widespread bleaching, which has severe repercussions for their health and their ability to recover from damage.

The Disruption of Ecosystems

The decrease of coral reefs has an effect on the entire marine food web since these reefs provide critical habitat for a wide variety of marine organisms. The biodiversity that corals help support is suffering along with their deterioration and loss of structure. When an ecosystem is disturbed, a wide variety of marine creatures suffer from a depletion of their available food supplies, habitats, and breeding grounds.

3. **Effects on Different Types of Shellfish**

Shellfish's Role in the Food Chain

Oysters, mussels, clams, and other bivalves, along with other types of shellfish, are essential components of marine ecosystems as well as human economies. They not only supply coastal communities with a source of protein and revenue through aquaculture and commercial fisheries, but they also function as filter feeders, which helps to improve the overall water quality.

Formation of a Shell

Shellfish are dependent on calcium carbonate for the formation of their shells, a process that is sensitive to changes in the chemical makeup of the ocean. Because of the progression of ocean acidification, there is less availability of carbonate ions (CO3-) for shell construction. This has an effect on the capacity of shellfish to construct and maintain their protective shells.

Possibility of being vulnerable

As a result of their calcifying nature, shellfish are particularly susceptible to the effects that ocean acidification can have. The larval and juvenile stages of shellfish are particularly vulnerable because a low pH might hamper their capacity to form shells, which in turn makes them more susceptible to predation and environmental stressors. Larvae and juveniles of shellfish are particularly at risk.

Consequences on a Commercial and Economic Scale

The shellfish industry and the economies of coastal areas face major challenges as a result of ocean acidification. Having fewer shellfish populations can result in having less opportunities to gather seafood, which can have a detrimental effect on the livelihoods of those who rely on these resources for their income and nourishment.

4. **Other Marine Organisms That Have Been Affected**
Planktonic algae known as phytoplankton

The fixation of carbon into the environment through photosynthesis is impossible without phytoplankton, which forms the basis of marine food networks. The process of ocean acidification has the potential to influence both the growth and the composition of communities of phytoplankton, which could have repercussions for both marine ecosystems and the climate system.

Fish

Ocean acidification can potentially have a knock-on effect on fish populations in a roundabout way. The supply of prey can be disrupted when there are shifts in lower trophic levels, such as phytoplankton and zooplankton, which can have an effect on the number of fish species as well as their distribution.

The Echinodermata

Calcifying species such as sea urchins and sea stars, which fall under the category of echinoderms, are susceptible to the effects of ocean acidification. The loss of these creatures, which play significant roles in marine ecosystems, has the potential to have a cascade influence on the dynamics of the ecosystem.

The cephalopods

Cephalopods, which include squid and cuttlefish, are particularly sensitive to shifts in the chemical composition of the water.

The acidity of the ocean can have an effect on their physiology and behavior, which may have repercussions for interactions between predators and prey as well as ecosystems.

5. **Strategies for Adaptation and Mitigation of Impacts**

It is necessary to implement both adaptation and mitigation techniques in order to protect marine life from the negative effects of ocean acidification.

Bringing Down Our CO2 Levels

The reduction of CO2 emissions at their points of origin is the most effective method for mitigating the negative effects of ocean acidification. Essential initiatives include making the switch to clean and renewable energy sources, improving energy efficiency, and decreasing dependency on fossil fuels.

Aquaculture that is not Unsustainable

It is absolutely necessary to encourage sustainable aquaculture practices that reduce the negative effects of shellfish farming on the surrounding environment. Monitoring the chemistry of the water, breeding animals selectively for resistance, and improving the circumstances in which shellfish are reared are all strategies.

Protected Marine Areas (or MPAs)

The creation of marine protected zones can be of assistance in the preservation of coral reefs and other delicate marine ecosystems. The marine life that lives in these protected regions has a better chance of thriving and adapting to changing conditions if these areas are allowed to function as refuges.

Investigation and Observation

Understanding the increasing problems and establishing successful adaptation strategies requires ongoing research and monitoring of ocean acidification and its repercussions. This is vital for the development of effective adaptation measures. The advancement of scientific research is critically important to both the broadening of our knowledge and the discovery of novel solutions.

The acidification of the ocean poses serious risks to marine life, particularly coral reefs and shellfish, which are particularly susceptible to shifts in the chemical makeup of the ocean. Understanding the effects of ocean acidification is crucial to the conservation of marine ecosystems and the resources they supply as CO_2 emissions from human activities continue to grow.

In order to ensure the long-term health and resiliency of marine life in the face of this global environmental crisis, it is essential to implement initiatives to reduce CO_2 emissions, promote sustainable aquaculture, develop marine protected areas, and invest in research and monitoring efforts.

7.3 Global consequences for fisheries and food security

The global fisheries industry plays a critical part in ensuring that people all around the world will have access to sufficient food supplies. The sustainability of fisheries, on the other hand, is coming under growing attack from a number of different sources, such as overfishing, the destruction of habitat, and climate change. In this essay, we will investigate the global repercussions of the situation for fisheries and food security, focusing on the difficulties they now confront as well as potential solutions to those difficulties.

1. **The Importance of Fishing Industries to the Nation's Food Supply**

   Source of Protein in the Diet

   Millions of people, particularly those living in coastal regions and communities with lower incomes, rely on fisheries as an important source of nutritional protein. Due to the abundance of critical nutrients that they provide, fish and other types of seafood are an essential part of the diet in many different locations.

   Food Supply Around the World

   The food supply on a global scale is significantly impacted by the fishing industry. The Food and Agriculture Organization (FAO) estimates that in 2018, fish accounted for more than 20 percentage points of the world's total supply of animal protein. Fish products are utilized in a wide variety of different processed foods, in addition to being consumed directly by humans.

   Earning a Living and Financial Support

   The fishing industry provides a means of subsistence and income for millions of people, including fishermen, those who work in fish processing, and those who are involved in the marketing and distribution of fish products. The fishing industry is a significant source of employment and cash in a great number of less developed countries.

2. **Obstacles Facing Fishing Industries Across the World**

   Excessive Fishing

   One of the most significant challenges facing the world's fisheries today is overfishing, which is defined as the unsustainable exploitation of fish stocks beyond their reproductive capability. It has the potential to reduce fish populations, wreak havoc on ecosystems, and threaten the economic viability of fishing enterprises over the long run.

Deterioration of Habitat

Fish populations and marine ecosystems can suffer as a result of habitat deterioration brought on by human activities such as coastal development, pollution, and bottom trawling, amongst other activities. It is necessary for fish to have healthy environments in order for them to spawn, feed, and hide.

Changes in Climate

Ocean temperatures, currents, and the geographic distribution of marine animals are all changing as a direct result of human-caused climate change. Because of the difficulty in accurately predicting and adjusting to these alterations, the availability and abundance of fish stocks may be negatively impacted as a result of these changes.

It's a Pollution

The contamination of marine ecosystems and the harm that it causes to fish populations can be attributed to pollution that is caused by agricultural runoff, industrial operations, and waste plastic. Seafood has the potential to accumulate contaminants, which poses a threat to human health.

IUU Fishing refers to fishing that is illegal, unreported, and unregulated

Illegal, unreported, and unregulated fishing (IUU fishing) is a worldwide issue that threatens the viability of fisheries. It reduces available fish stocks, poses a risk to biodiversity, and causes market disruption by providing unfair competition to fishermen who are working lawfully.

3. The Repercussions for Fisheries Around the World

Fall in the Amount of Fish Stocks

The decrease of many fish stocks around the world can be attributed to both excessive fishing and the destruction of their natural habitats.

The extinction of important species has the potential to disrupt the balance and productivity of ecosystems, resulting in a diminished supply of fish for human consumption and commercial commerce.

Losses to the Economy

It is possible for the fishing industry to suffer economic setbacks due to dwindling catches, rising operating expenses, and disruptions brought on by illegal fishing techniques and excessive fishing overall. This, in turn, has an impact on the means of subsistence and income of those whose livelihoods depend on the sector.

Instability of Food Supply

A decrease in the availability of fish can contribute to food insecurity, particularly in regions where fish is a key source of protein and critical nutrients. This is especially true in regions where fish is a primary source of protein. There is a danger of malnutrition in vulnerable populations, such as coastal villages and households with low incomes.

Effects on One's Health

The decrease in the consumption of fish may have unfavorable effects on health due to the fact that fish are a good source of critical nutrients such as omega-3 fatty acids, vitamin D, and high-quality protein. A lower consumption of fish could be a factor in the development of diet-related health problems, such as malnutrition and non-communicable diseases.

Disruption on the Social and Cultural Level

Communities that rely on fishing frequently have strong cultural and social ties to the ocean. It is possible for declining fish stocks to cause social and cultural upheaval, with people suffering from the loss of their traditional means of subsistence and ways of life.

4. **Methods for Maintaining Sustainable Fisheries and Ensuring Food Safety**

Management that Takes Into Account Ecosystems

Adopting management strategies that are based on ecosystems can assist in the upkeep of healthy marine ecosystems and ensure the continued viability of fisheries over the long run. These methods take into account the interdependencies that exist between different species, ecosystems, and ecological processes.

Methods of Fishing That Are Sustainable

It is possible to assist avoid overfishing and safeguard fish stocks from depletion by promoting sustainable fishing methods. Some examples of these measures are catch quotas, seasonal closures, and gear restrictions. Legal rules and international agreements are two avenues that might be pursued in order to put these measures into effect.

Protection and Rehabilitation of Natural Habitats

It is absolutely necessary to protect and restore important marine ecosystems in order to keep fish populations stable. Some examples of these habitats include coral reefs, seagrass beds, and mangroves. Marine protected areas and environmentally responsible coastal development are two examples of possible protective measures.

Taking Action Against Climate Change

It is imperative for the future of fisheries that measures be taken to both mitigate the consequences of climate change and adapt to them. It is possible to lessen the negative effects of climate change on fish supplies by lowering emissions of greenhouse gases, investing in renewable energy sources, and putting into practice climate-resilient activities.

Efforts Made to Fight IUU Fishing

In order to combat illegal, unreported, and unregulated fishing (IUU fishing), efforts are being made to promote international collaboration, improve monitoring and surveillance systems, and create traceability methods for marine products. These precautions have the potential to mitigate the damaging effects of unlawful fishing.

Aquaculture that is not Unsustainable

Aquaculture, often known as the farming of fish and other seafood, has the potential to supplement natural fisheries and assist in meeting the rising demand for fish products around the world. It is vital to implement sustainable aquaculture methods, such as those that involve responsible feed management and minimized environmental impacts.

Management that is based on the community

Participation of local people in fisheries management has the potential to result in more environmentally responsible actions. Collaborative effort, the preservation of resources, and the defense of mutual interests are all fostered by community-based management.

5. **Agreements and Initiatives Concluded at the International Level**

The Sustainable Development Goals (SDGs) set forth by the United Nations

Through the framework of its Sustainable Development Goals, the United Nations has acknowledged the significance of fisheries and the protection of food supplies. Goal 14, titled "Life Below Water," places a particular emphasis on the protection of marine resources and the responsible management of their usage. The successful completion of this objective will contribute to the preservation of marine ecosystems as well as the safety of our food supply.

The Code of Conduct for Responsible Fisheries Developed by the FAO

The Food and Agriculture Organization of the United Nations (FAO) is the organization that is responsible for developing the Code of Conduct for Responsible Fisheries. It provides recommendations and principles for managing sustainable fisheries, addressing concerns such as overfishing, illegal, unreported, and unregulated fishing (IUU), as well as protecting habitats.

Organizations for Regional Fisheries Management (also known as RFMOs)

International organizations known as RFMOs are responsible for the management and regulation of fisheries in particular maritime zones. They make the guidelines and take the precautions necessary to ensure the responsible consumption of fish stocks. It is impossible to successfully manage fisheries without the collaboration of multiple nations working through RFMOs.

The global fisheries industry and people's access to food are both suffering severely as a direct result of dwindling fish stocks and fishing practices that are not sustainable. In order to effectively address these difficulties, a comprehensive strategy that takes into account a variety of factors—including sustainable management, habitat protection, climate change mitigation, and community engagement—is required. There is a significant need for international agreements and initiatives to play a part in the promotion of responsible fishing methods and the protection of the health of coastal people, ecosystems, and food security on a global scale. It is not just necessary for the health of millions of people who rely on fish as their major source of food and income that we ensure the sustainability of fisheries; it is also an urgent requirement in order to meet our obligations to the environment.

Chapter 8

Extreme Weather Events

Hurricanes, heatwaves, droughts, floods, and wildfires are just some of the extreme weather phenomena that have become more frequent and severe in recent years. As a result, communities all over the world, as well as economies and ecosystems everywhere, face tremendous difficulties. These occurrences, which are frequently made worse by changes in climate, have repercussions that are far-reaching for the environment, agriculture, and public health alike. In this in-depth study, we will investigate the nature of extreme weather events, as well as their origins and repercussions, as well as techniques for minimizing those effects on a global scale as well as for adapting to them on a local one.

1. **Gaining an Understanding of Severe Weather Phenomena**
   The Concept and Its Variations
   The deviation of a weather event from its normal pattern, which manifests as a phenomenon that is abnormally intense, lengthy, or destructive, is the defining characteristic of an extreme weather event. Hurricanes, tornadoes, heatwaves, coldwaves, droughts, floods, and wildfires are all examples of common types of extreme

weather, and each has its own set of features as well as an influence on a separate region.

The effects of climate change and severe weather

The extent to which climate change is affecting the occurrence of extreme weather is an important topic of research. The rise in average global temperature has been related to an increase in the frequency and severity of certain weather phenomena, including heatwaves and violent storms. This has led to an increase in the number of instances of extreme events occurring all over the world.

Both natural and anthropogenic factors are included in this study

Although natural variability does play a part in the occurrence of extreme weather events, human activities, such as the production of greenhouse gases and changes in land use, have increased the frequency of these occurrences as well as their intensity, contributing to a climate that is more unstable.

2. **Impact of Severe Storms and Other Extreme Weather**

Damage to the Infrastructure

Weather events of extreme severity are capable of causing substantial damage to a variety of infrastructure, such as buildings, roads, bridges, and utilities. Particular natural disasters, such as hurricanes, floods, and wildfires, have the potential to cause significant damage to vital infrastructure, thereby upsetting both people and economies.

Agriculture and the Securing of Food Supplies

It is possible for extreme weather events such as droughts, heatwaves, and floods to have a significant negative impact on agricultural output. This can result in failed crops, lost livestock, and diminished food supplies. Food insecurity and economic instability are potential threats to economically vulnerable regions, particularly those that are highly dependent on agriculture.

Dangers to the Public's Health

Hurricanes, heatwaves, and floods are all potentially serious threats to public health, as they can cause injuries, illnesses, and the spread of diseases that are transmitted by water and vectors. The displacement of communities, as well as disruptions to healthcare infrastructure, can make these risks even more severe, particularly for groups who are already vulnerable.

The Repercussions on the Environment

It is possible for extreme weather events to have substantial effects on the environment, including the destruction of habitats, the loss of biodiversity, and disruptions to ecosystems. Flooding, wildfires, and storms are all potential causes of soil erosion, water contamination, and the destruction of natural habitats, all of which pose a threat to the delicate balance that exists within local and regional ecosystems.

3. Frequent Categories of Severe Weather Occurrences

Cyclones such as hurricanes and typhoons

Powerful storm systems known as tropical cyclones, or hurricanes or typhoons depending on their location, are characterized by high winds, heavy rainfall, and storm surges. Tropical cyclones are also known as typhoons. These occurrences have the potential to cause considerable damage to coastal regions, which can lead to flooding, the destruction of infrastructure, and even fatalities.

Temperature Extremes

Heatwaves are stretches of time in which temperatures remain abnormally high for days or even weeks at a time. They can have major repercussions for human health, including the onset of heat-related illnesses and even fatalities, particularly among vulnerable populations and in urban areas with inadequate cooling infrastructure.

Continual dry spells

A protracted period of exceptionally low precipitation can cause droughts, which in turn can lead to a lack of available water, the failure of crops, and economic losses. They have the potential to have a considerable impact on agricultural practices, water supplies, and ecosystems, all of which can contribute to an increased risk of hunger and deterioration of the natural environment.

Continual flooding

Floods can be caused by excessive rainfall, storm surges, or the overflow of rivers and coastal areas. These are the three main contributing factors. They have the potential to cause widespread damage to infrastructure, the uprooting of communities, and the spread of waterborne diseases, all of which present dangers to public health and safety.

Forest fires

Wildfires are fires that are not under human control and can spread quickly across vegetation and woods. They are frequently made worse by dry weather and high temperatures. They can result in the destruction of natural habitats, the contamination of the air, as well as the loss of life and property, which creates obstacles for the rehabilitation of ecosystems and the resiliency of communities.

III. Strategies for Adaptation and Mitigation of Impacts
Reduced Impacts of Climate Change

It is absolutely necessary to cut greenhouse gas emissions by increasing the use of renewable energy sources, energy efficiency measures, and sustainable land-use practices. Doing so is the single most important thing that can be done to reduce the negative effects of extreme weather events. When it comes to addressing the underlying causes of climate change, international collaboration and policies that aim to reduce emissions are absolutely necessary.

Preparedness for Emergencies and Early Warning System Measures

It is possible for communities and governments to better predict and respond to extreme weather occurrences by making investments in comprehensive disaster planning and early warning systems.

This includes the creation of preparations for evacuation, the installation of emergency shelters, and the broadcast of information that is both timely and accurate to the general population.

Strong and Dependable Infrastructure

It is necessary to construct and maintain infrastructure that is robust and able to endure the effects of extreme weather events in order to lessen the amount of damage that occurs and to ensure that essential services are not interrupted. This involves the development of flood barriers, buildings with reinforced foundations, and infrastructure that is capable of withstanding severe winds and seismic activity.

Restoring Ecosystem Function While Protecting It

Protecting and restoring ecosystems, such as forests, wetlands, and mangroves, can help lessen the effects of severe weather events such as hurricanes and tornadoes. These natural buffers have the potential to lessen the effects of flooding, stabilize the soil, and provide habitats for various types of species.

Management of Water Resources That Is Sustainable

It is possible to lessen the impact of drought and water shortages by putting into practice sustainable water management strategies. These practices can include making more effective use of available water resources and cultivating crops that are resistant to drought.

Readiness with regard to Public Health

The effects of heatwaves, floods, and the spread of illnesses can be mitigated by strengthening public health systems and raising awareness of the health concerns connected with extreme weather events.

Education and a Consciousness Campaign

In order to create resiliency and encourage mitigation actions, it is vital to increase public awareness about the causes and consequences of extreme weather events, as well as the importance of individual and communal action.

Because of climate change and other anthropogenic reasons, extreme weather events are occurring more frequently and with greater intensity. These occurrences have repercussions that are far-reaching, particularly in the areas of agriculture, public health, and the environment. It is crucial to implement policies for mitigation and adaptation in order to handle these difficulties. Other important steps include reducing emissions of greenhouse gases, investing in infrastructure that is resilient, protecting ecosystems, and improving public health readiness.

The mitigation of the underlying causes of climate change and the development of community resilience are two essential actions that must be taken in order to reduce the severity of the effects of extreme weather events and protect the health and safety of people all over the world.

8.1 Examination of the relationship between global warming and extreme weather

The human influence on climate change has contributed to global warming, which in turn has led to a rise in the frequency and severity of extreme weather events across the globe. The impact of these catastrophes on communities, ecosystems, and economies has been significant. Some of these occurrences include intense hurricanes and heatwaves, while others include disastrous floods and wildfires. The purpose of this in-depth study is to investigate the complex connection that exists between rising global temperatures and extreme weather patterns. More specifically, the investigation will focus on the underlying mechanisms, the scientific evidence that supports this connection, as well as the consequences for future climate forecasts and adaptation methods.

1. **Gaining an Understanding of Climate Change and Global Warming**

   The Emission of Greenhouse Gases

   Carbon dioxide (CO_2), methane (CH_4), and nitrous oxide (N_2O) are examples of greenhouse gases that can be released into the atmosphere as a result of activities such as the burning of fossil fuels, the clearing of forests, and the usage of industrial processes. The greenhouse effect describes the process by which these gases retain heat, which in turn causes an increase in the average temperature of the earth.

   The Dynamics of Climate Change

   The buildup of greenhouse gases in the atmosphere causes a

disruption in the climate system of the Earth, which in turn causes shifts in temperature, changes in precipitation patterns, shifts in sea level, and variations in the weather. This disturbance is a primary factor in the observed shifts in worldwide weather patterns as well as the rise in the frequency of extreme weather events.

2. **The Different Forms of Severe Weather and Their Identifiable Traits**

Cyclones such as hurricanes and typhoons

Hurricanes and typhoons are both types of tropical cyclones, which are violent storm systems that are characterized by strong winds, torrential rainfall, and storm surges.

The name of the storm system changes based on its location. They have the potential to inflict major damage on coastal regions, which can lead to extensive flooding and the destruction of infrastructure.

Temperature Extremes

Heatwaves are extended episodes of extremely high temperatures that can have a significant negative impact on human health, agriculture, and ecosystems. Heatwaves can be caused by a variety of factors. In metropolitan regions that do not have enough cooling infrastructure, they might cause heat-related illnesses, crop failures, and water shortages. This is especially true in the case of urban areas.

Continual dry spells

Water scarcity, decreased agricultural output, and ecological disruptions can all be the direct results of prolonged periods of exceptionally low precipitation, which is what we refer to as droughts. They have the potential to bring about problems with the availability of food and water, as well as contribute to the destruction of natural habitats and ecosystems.

Continual flooding

An excessive amount of rainfall, storm surges, or the overflow of rivers and coastal areas can all contribute to the inundation of land, which is what causes floods. They are capable of causing major damage to infrastructure, the relocation of people, and the spread of waterborne diseases, all of which pose threats to the public's health and safety.

Forest fires

Wildfires that are not contained can quickly spread across forests and other vegetation, which can result in the destruction of natural ecosystems, the polluting of the air, as well as the loss of life and property. Wildfires are typically made worse by dry weather and high temperatures.

3. **The Evidence That Shows There Is a Connection Between Global Warming and Extreme Weather**

Data Derived From Observations

Long-term observational data from a variety of meteorological and climatic monitoring systems give persuasive evidence that there is a correlation between global warming and the increase in frequency and intensity of extreme weather events. The data presented here highlight the patterns that have been observed over the past century in terms of variations in the distribution of precipitation, changes in sea level, and increases in average temperatures.

Modeling of the Climate

Complex climate models, which were created through the application of sophisticated computer methods, anticipate potential future climatic scenarios based on a variety of emission paths for greenhouse gases. These models forecast an increase in the occurrence of extreme weather events, which provides vital insights into the possible implications of prolonged global warming on regional as well as global scales.

The General Agreement of Scientists

The scientific world has formed a consensus, as a result of significant study and papers that have been peer-reviewed, about the connection between rising global temperatures and an increase in the frequency and severity of extreme weather events. The Intergovernmental Panel on Climate Change (IPCC) and a large number of scientific organizations from around the world have stressed the critical importance of addressing climate change as quickly as possible in order to reduce the risks associated with it.

4. **The Mechanisms That Are Amplifying The Impact Of Global Warming On The Weather**

The Warming of Both the Atmosphere and the Oceans

Storms, hurricanes, and typhoons are able to generate a greater amount of destructive force as a direct result of global warming's contribution to an overall rise in atmospheric and oceanic temperatures. Warmer waters can also contribute to an increase in the intensity of precipitation, as well as the generation of storms that are more intense and carry more moisture.

Polar Ice Caps Are Melting

The result of rising global temperatures, the melting of polar ice caps and glaciers adds to the rise in sea levels, which in turn increases the risk of coastal floods and storm surges. This phenomenon is a consequence of global warming. Alterations in the climate systems of the entire planet are a direct result of disruptions in atmospheric circulation patterns brought about by changes in the Arctic and Antarctic regions.

Alterations to the Circulation of the Atmosphere

The patterns of atmospheric circulation can be influenced by global warming, which can then lead to changes in the location of jet streams, the number of weather fronts, and the distribution of precipitation. These changes can lead to extended periods of drought, heatwaves, and an increase in the intensity of extreme weather events in particular places.

5. **Implications for Ecosystems and the Variety of Life on Earth**

The Disruption of Ecosystems

The disruption of ecosystems and natural habitats by extreme weather can result in the loss of biodiversity, changes in the distribution of species, and the deterioration of important ecosystems including forests, wetlands, and coral reefs. These disturbances have the potential to have long-lasting effects on the ecosystem services provided by natural habitats as well as their resilience.

Loss of Habitat and Fragmentation of Habitat

The worsening of extreme weather events can result in the destruction and fragmentation of habitats, which in turn threatens the continued existence of a great number of species as well as their capacity to adjust to shifting climatic conditions. When habitats are degraded, the animals that live there may lose access to essential breeding grounds, food sources, and migration routes as a result.

Threats to the Survival of Species and Their Potential for Extinction

Species that are sensitive to climate change, including those that have specific habitat requirements and temperature tolerances that are on the lower end, are especially susceptible to the negative effects of extreme weather events. These occurrences can raise the chances of a species going extinct, cause disruptions in food chains, and lead to impacts that cascade throughout entire ecosystems.

6. **Methods for Lessening the Effects of Severe Storms and Other Extreme Weather Events**

The Lowering of Carbon Emissions

It is absolutely necessary to cut carbon emissions by increasing the use of renewable energy sources, energy efficiency measures, and sustainable land-use practices if we are going to lessen the effects of global warming and cut down on the number and severity of extreme weather occurrences.

Planning for both Adaptation and Resilience

Communities and regions can be helped to better cope with the effects of extreme weather events by developing strong adaptation and resilience plans that combine climate-responsive infrastructure, early warning systems, and community-based disaster preparedness strategies.

Management of Land and Water That Is Sustainable

The implementation of sustainable land and water management strategies, such as reforestation, soil conservation, and water resource management, can help reduce the hazards of soil erosion, flooding, and water scarcity, thereby enhancing the resilience of ecosystems and the sustainability of resources.

The Protection and Regeneration of Biological Diversity

Protecting and restoring ecosystems that are rich in biodiversity, such as forests, wetlands, and coastal habitats, is essential for sustaining natural resilience and enabling the adaptation of species to changing climatic conditions. This includes the protection and restoration of forest ecosystems.

Cooperation on a Global Scale and Various Policy Frameworks

For the purposes of combating global warming, fostering sustainable development, and constructing global resilience to extreme weather events, it is essential to have international cooperation and the development of comprehensive policy frameworks, such as the Paris Agreement and the United Nations Sustainable Development Goals (SDGs).

The relationship between rising global temperatures and an increase in the frequency and severity of extreme weather events is complex, and there is scientific evidence to support the link between rising global temperatures and this increase in both frequency and intensity. The warming of the atmosphere and seas, the melting of polar ice, and changes in air circulation are the mechanisms that contribute to an increase in extreme weather as a result of global warming. The fact that the implications of this link extend to ecosystems, biodiversity, and communities who are particularly susceptible highlights how urgent it is to address climate change and create resilience to extreme weather.

It is absolutely necessary to implement mitigation techniques, such as cutting down on carbon emissions and practicing environmentally responsible land and water management, in order to slow the progression of global warming and lessen the damage caused by extreme weather. Planning for adaptation and resilience, together with the protection of biodiversity and the collaboration of nations around the world, are essential elements of an all-encompassing strategy for addressing this complex and multifaceted challenge. In light of the findings of case studies and observations made in real life, it is clear that preventative actions are required to shield communities, ecosystems, and the global environment from the growing dangers posed by extreme weather events as a result of global warming.

8.2 Climate adaptation and disaster management

Changes in global weather patterns and a rise in the frequency and intensity of natural disasters are being caused by climate change, which is being driven by anthropogenic factors such as emissions of greenhouse gases, deforestation, and activities in the industrial sector. The world's civilizations, ecosystems, and economy are all going to face substantial challenges as a result of these changes. As a result of these dangers, the implementation of policies for climate adaptation and catastrophe management has become absolutely necessary in order to construct resilience in the face of a changing climate. This in-depth study investigates the fundamental ideas, difficulties, and solutions linked with the management of climate change and natural disasters.

1. **Acquiring Knowledge about Climate Change Adaptation The Meaning and the Aim of It**

   The term "climate adaptation" refers to the process of adjusting to or preparing for changes in the climate, with the ultimate goal of minimizing the vulnerabilities and dangers associated with these changes. It comprises a wide variety of different measures and policies that are designed to safeguard populations, ecosystems, and infrastructure against the effects of climate change.

The Foundations of Climate Change Adaptation

The ideas of foresight, adaptability, capacity building, and flexibility are fundamental to the practice of climate adaptation. The identification of potential climate hazards should be the first emphasis of adaptation strategies, followed by the modification of systems and behaviors, the construction of adaptive capacity, and the maintenance of flexibility to respond to changing circumstances.

2. **The Difficulty Presented by Climate Change**

The Effects of Climate Change

The effects of climate change are manifested in a variety of ways, including an increase in temperature, a shift in the patterns of precipitation, an increase in the frequency and severity of meteorological events, a rise in sea level, and an acidity of the ocean.

The agriculture industry, water supplies, human health, ecosystems, and economies will all be significantly impacted as a result of these changes.

Populations at Risk or in Danger

The effects of climate change are frequently felt most acutely by vulnerable populations, such as low-income communities, indigenous peoples, and other groups that have historically been neglected. Inequalities in social and economic standing can make the difficulties of adapting to changing conditions and managing disasters even more difficult.

Consequences on a Global Scale

The subject of climate change affects people all across the world and has far-reaching repercussions. For instance, if sea levels continue to rise, this could force millions of people living in coastal areas to relocate, which would wreak havoc on the region's social, economic, and political systems. The global community has a responsibility to address the linked nature of climate change and the effects it has.

3. **Essential Ideas in Climate Change Adaptation**
   Capacity for Bouncing Back
   The ability of a system, whether it be human, ecological, or infrastructure-based, to recover from disturbances, adapt to new circumstances, and withstand shocks is what is meant by the term resilience. Increasing resiliency is one of the primary goals of adaptation attempts to climate change.
   Adaptation Methods and Techniques
   Strategies for climate adaptation comprise a wide variety of different measures, ranging from structural solutions like building stronger infrastructure to non-structural techniques like fostering ecosystem-based adaptation and improving early warning systems. Examples of structural solutions include creating stronger infrastructure.
   Climate Change Legitimacy
   The concept of climate justice places an emphasis on the moral and equitable distribution of the resources and benefits associated with climate adaptation and disaster management. This helps to ensure that the most vulnerable communities are not disproportionately burdened by the effects of climate change.

4. **Obstacles to Climate Change Adaptation**
   Vacancies in Our Knowledge and Information
   Inadequate research, data, and knowledge regarding the local and regional impacts of climate change can be a barrier to the development of efficient adaptation strategies. Filling up the gaps in one's knowledge is an essential step in developing correct and evidence-based tactics.
   Constraints Placed on Resources and Finances
   When it comes to putting adaptation measures into action, a number of nations and communities, particularly those with restricted access to resources, are hampered by financial and capability constraints. It is absolutely necessary to have adequate financing and support in order to achieve climate resilience.

Coordination and Administrative Oversight

It is essential for there to be strong cooperation between the many government agencies, non-governmental groups, and communities in order to successfully adapt to climate change. The long-term aims of climate resilience need to be aligned with the governance structures and policies that are in place.

5. **Strategies for Climate Change Adaptation**

Resilience of the Infrastructure

It is very necessary to adapt infrastructure, such as buildings, transportation systems, and energy networks, to withstand the effects of climate change, such as more powerful storms, heatwaves, and rising sea levels. Doing so is critical for preserving key services and preventing economic disruptions.

Adaptation Based on Ecosystems (EBE)

Adaptation strategies that are ecosystem-based involve making use of natural systems, such as wetlands and forests, in order to lessen the negative effects of climate change. These ecosystems can be better managed, along with water supplies and the risk of flooding, if we preserve and restore them. This will also help support biodiversity.

Systematic Methods of Early Warning

Improving early warning systems, which should include monitoring of meteorological and hydrological conditions, is essential for giving communities with timely knowledge about imminent climate-related disasters, which in turn enables people to prepare for the disasters and evacuate if necessary.

Participation in the Community

It is crucial for the process of building resilience to incorporate local people in the process of designing and putting into action adaption measures. Knowledge and perspectives gained at the local level can be beneficial to decision-making and help ensure that tactics are suitable for the setting in question.

Agriculture that is not affected by climate change

Adapting agricultural practices to ever-shifting weather patterns is an essential step toward achieving and maintaining food security. This can be accomplished through the use of measures such as crop variety, soil conservation, and farming methods that are water-efficient.

Preparedness in the Areas of Health and Public Health

In order to handle issues such as heat-related illnesses, the expansion of vector-borne diseases, and the requirement for disaster medical response skills, public health systems need to adapt to the changing environment.

6. **Understanding Emergency Preparedness and Response**

The Framework for Disaster Management

Mitigation, preparedness, reaction, and recovery are the four stages that are included in the process of disaster management, which is a methodical procedure. It includes things like evaluating risks, making plans, allocating resources, and coordinating a reaction, among other things.

Types of Catastrophes

Natural disasters can include earthquakes, hurricanes, and floods; man-made disasters can include industrial accidents, nuclear events, and conflict-related crises; and natural disasters can include earthquakes, hurricanes, and floods. Extreme weather that is influenced by climate change typically plays a role in the occurrence of natural disasters.

7. **An Integrated Approach to Climate Adaptation and Disaster Management**

Challenges That Are Interconnected

The management of natural disasters and climate change adaptation are inextricably linked. The integration of these two sectors is required in order to build resilience to the effects of climate change because the frequency and severity of natural catastrophes associated to climate change are expected to grow.

Risk Avoidance and Emergency Preparation

The reduction of vulnerabilities, the enhancement of capacity to respond to catastrophes, and the facilitation of recovery are the three primary goals of an integrated approach, which places a primary emphasis on preparedness and risk reduction. The incorporation of climate adaption techniques into catastrophe management strategies is strongly recommended.

Capacity for Adaptation

The capacity of communities and organizations to change to shifting conditions and enhance their ability to respond to natural disasters is referred to as their adaptive capacity. Increasing one's capacity for adaptation is a crucial component of an approach that is integrated.

8. Obstacles to Overcome and Opportunities to Seize in the Future

Financial Means Available

The acquisition of sufficient funds and resources presents a significant obstacle for those working in the fields of climate adaptation and catastrophe management. These activities demand a large investment in building capacity, early warning systems, and infrastructure.

Cooperative efforts on a global scale

The effects of climate change can be felt across national boundaries. When it comes to addressing global difficulties and fostering a coordinated response to climate adaptation and catastrophe management, international cooperation and agreements like the Paris Agreement play a significant role in addressing global challenges and achieving these goals.

Research and development both

Putting money into research and innovation can result in the development of adaptation and crisis management strategies that are both more effective and more environmentally friendly. The application of recently developed methodologies and tools is one way to address newly arising problems and possibilities.

Building resilience in the face of a changing climate requires incorporating a number of different strategies, including climate adaptation and catastrophe management. An integrated approach that focuses on decreasing risks, boosting readiness, and building adaptive capacity is required because of the linked problems posed by the impacts of climate change and the rising frequency of climate-related disasters. This approach is necessary because of the interwoven nature of these challenges.

To find solutions to these problems, national governments, local communities, and international organizations need to put money, research, and collaboration at the top of their agendas. An effective response to the effects of climate change and natural disasters requires the coordinated efforts of multiple stakeholders, innovative solutions, and a commitment to climate justice. This will ensure that the most vulnerable populations are not left behind in the pursuit of building resilience for a sustainable future.

Chapter 9

Socioeconomic Implications

The current state of the climate is a global emergency that has significant repercussions for society and the economy. It is having an effect on everything from economic stability and public health to food security and migratory patterns, and it is transforming the way societies function as a result. This in-depth study investigates the complex connection that exists between climatic shifts and the repercussions they have on society and the economy. Particular attention is paid to the difficulties that are being experienced by vulnerable populations as well as the urgent requirement for adaptive and mitigating solutions to deal with the effects of climate change.

1. **The Existing Socioeconomic Conditions**
   What Do We Mean When We Talk About Socioeconomic Implications?
   The effects of climate change on societies, economies, and the general well-being of individuals and groups are what are referred to as the socioeconomic implications of climate change. These repercussions present themselves in a variety of ways, including

upheavals in people's means of subsistence, adjustments in their patterns of production and consumption, and variations in their access to vital resources.

The Problem of Climate Change from a Number of Different Angles

The problem of climate change is not merely one of environmental concern; rather, it is a complicated and multifaceted problem that has significant repercussions for both society and economies. For the purpose of making informed decisions and developing effective policy, it is vital to have a solid understanding of the socioeconomic implications of climate change.

2. The Effects of Climate Change on Economic Structures

Disruption to the Economy

The increasing expenses associated with damage to infrastructure, insurance payouts, and the requirement for disaster response and recovery are all caused by climate change, which causes economic systems to become unstable. There is a potential for significant economic damage to result from extreme weather events, rising sea levels, and shifting precipitation patterns.

Inequality as well as precariousness

The effects of climate change are making inequality worse because they are having a disproportionately negative impact on vulnerable people, who frequently lack the means and adaptive capacity to deal with the effects of climate change. Communities with low incomes and those that are already marginalized have a greater threat of negative socioeconomic outcomes.

The effect on the GDP

Both direct economic losses and indirect effects on industries such as agriculture, tourism, and healthcare can have a substantial impact on the overall Gross Domestic Product (GDP) of countries. Direct economic losses can be caused by climate change. Indirect effects can be caused by climate change. These repercussions will have a trickle-down effect on economic growth as well as jobs.

3. **Agriculture and the Safety of the Food Supply**
Agricultural Practices That Are Continually Evolving
The effects of climate change on agricultural practices are already being seen, with decreased crop yields, increased risks to food production, and hampered distribution networks. The rise in average temperatures, varying patterns of precipitation, and proliferation of both pests and illnesses are all factors that contribute to these transitions.

The Unpredictability of Food Prices
The unpredictable weather patterns and catastrophic occurrences can cause fluctuations in the price of food, which can make it difficult for disadvantaged communities to obtain food that is both inexpensive and healthy. This volatility may help to cause social unrest as well as economic instability.

The Influence on People's Ability to Make a Living
Agriculture provides a means of subsistence for literally billions of people all over the world. Challenges to food security and agriculture brought on by climate change can bring about decreased income, forced relocation, and an increased dependency on outside assistance for communities who are adversely affected.

4. **Health Care for the Public**
Illnesses Associated with Heat
The likelihood of developing heat-related illnesses, such as heat exhaustion and heat stroke, increases as temperatures continue to rise. Vulnerable people, such as the elderly and those who already have one or more preexisting medical issues, are at a greater risk.

Diseases Transmitted by Vectors
The distribution of disease vectors, such as mosquitoes that spread diseases such as malaria, dengue, and Zika, is impacted by climate change. There is a potential for an increase in the spread of these illnesses if the vectors' ranges continue to expand.

Quality of the Air
Alterations in temperature and weather patterns can aggravate

existing respiratory disorders and make existing air pollution even more severe. Low air quality has been related to a variety of health concerns, including those affecting the respiratory system and the cardiovascular system.

5. **Water Supply and Distribution**
Dry conditions and a lack of available water
The most significant effects of climate change are the droughts and water shortages that have severe repercussions for the agricultural industry, as well as for the provision of drinking water and industrial activities. These occurrences have the potential to spark contention over the limited water resources.

Inundation as well as Diseases Transmitted by Water
On the other hand, excessive rainfall and flooding have the potential to contaminate water supplies, which in turn raises the danger of diseases that are transmitted by water. In the aftermath of floods, getting access to water that is both clean and safe becomes a key concern.

Production of Hydroelectricity
The availability and dependability of hydropower generation could be impacted by climate change, which would have repercussions for the energy industry and, as a consequence, for the stability and growth of the economy.

6. **Movements of People and Displacements**
Migration That Is Caused By Climate Change
Some of the socioeconomic repercussions of climate change include migration and displacement of people against their will. People may be forced to abandon their homes and look for safety in other parts of the world due to factors like as rising sea levels, extreme weather, and a lack of available resources.

Pressure from Urbanization on Existing Infrastructure
Migrants who are forced to relocate due to climate change frequently end up in urban areas. This inflow has the potential to put a strain on infrastructure, services, and resources, which

may in turn lead to socioeconomic issues for both the migrant population and the communities that are hosting them.

Conflict as well as Unrest in Society

In parts of the world where climate change is wreaking havoc, competition for scarce resources like land and water may easily escalate into armed conflict and societal upheaval. The disruption of governance, economies, and people's means of subsistence is one of the socioeconomic effects that these consequences have over the long term.

7. **Strategies for Adaptation and Mitigation of Impacts**

Climate-Resilient Adaptation

Increasing a community's ability to withstand the effects of climate change is very necessary in order to address its socioeconomic effects. This entails modifying infrastructure, agricultural practices, and public health systems such that they are resistant to the effects of a changing climate.

Renewable Sources of Energy

It is absolutely necessary to make the switch to renewable energy sources like solar and wind power in order to reduce the negative effects that climate change will have. The trend of global warming and the related repercussions can be slowed down by taking steps to reduce emissions of greenhouse gases.

Preparedness for Emergencies

The socioeconomic effects of climate-related catastrophes can be mitigated somewhat by employing measures that are effective in disaster preparedness and response.

Important aspects include early warning systems, evacuation preparations, and active participation from the community.

Agriculture that is Environmentally Sound

Food insecurity is a problem that can be helped and farming communities' livelihoods can be supported by using sustainable agriculture methods. These practices can include crop diversification and techniques that are water efficient.

Climate Change Legitimacy

In the context of climate change, climate justice refers to the equal allocation of resources and benefits, with the goal of ensuring that the most vulnerable groups are not unfairly burdened by the socioeconomic effects of climate change. This strategy places a strong emphasis on fairness in both the adaptation and mitigation activities that are taken.

The socioeconomic repercussions of climate change are extremely extensive, having an impact not only on economies but also on public health and the safety of food supplies. Vulnerable communities are hit harder than other populations, which compounds the inequality that already exists. In order to address these issues and construct a more resilient and just future, effective adaptation and mitigation methods are required. It is impossible to exaggerate the urgent need for global cooperation, equitable resource allocation, and innovative solutions at a time when societies all over the world are struggling to come to terms with the complicated and ever-changing effects of climate change.

9.1 The human toll of global warming, including displacement and health effects

The rising concentrations of greenhouse gases in the atmosphere are the primary driver of global warming, which has far-reaching implications not just for the natural world but also for human communities and economy. One of the most significant and immediate repercussions is the toll it takes on people's lives, especially in the form of relocation and bad health effects. This is one of the most serious and immediate consequences. As the average temperature of the planet continues to rise and as the frequency and severity of extreme weather events continues to increase, many people and communities will be forced out of their homes and will suffer a variety of health-related difficulties. This exhaustive study investigates the human toll of global warming, providing light on the relocation of vulnerable groups as well as the health repercussions that accompany changes that are caused by climate change.

1. **Movement and relocation As a result of Climate Change**
   What it is, and what causes it
   When discussing climate change within the context of global warming, the term "displacement" refers to the forcible removal of people from their homes and communities as a direct result of the adverse effects of climate change. The rise in sea level, the occurrence of catastrophic weather events, and gradual shifts in climatic circumstances are some of the reasons that contribute to relocation.

   Rising Seas and Coastal Relocation Due to Climate Change
   A direct result of global warming is an increase in the level of the ocean. The melting of polar ice and the expansion of ocean waters contribute to an increase in global sea levels as the temperature of the Earth continues to rise. This phenomena has a disproportionately negative impact on low-lying coastal areas, causing millions of people to be forced out of their homes and into temporary housing.

   Extremely Harsh Weather Occurrences
   As a direct result of global warming, catastrophic weather occurrences, such as hurricanes, floods, and wildfires, are occurring with greater frequency and intensity. As a result of these events, entire communities may be uprooted all of a sudden because they are required to flee in the face of an impending disaster.

   Displacement With a Slow Start
   Alterations in the environment, such as those brought on by protracted droughts, the spread of desertification, and the depletion of freshwater supplies, can also bring about population movement over time. It's possible that people will have to abandon their homes if the climate continues to change in such a way that makes them inhabitable.

2. **The Effects of Climate-Induced Displacement on People's Health**
   Impacts on a Person's Physical Health

When people are displaced, they frequently experience negative effects on their physical health. These can include being exposed to severe weather conditions, an increased risk of injuries when evacuating, and the spread of waterborne infections in temporary shelters that are overcrowded.

Problems Relating to Mental Health

The emotional toll that can be taken by being uprooted can be enormous, and it can lead to difficulties with mental health such as anxiety, depression, post-traumatic stress disorder (PTSD), and the feeling of not belonging or having security in one's life.

Access to Medical Care

There may be barriers to accessing healthcare for displaced populations, including as interruptions to medical services, the loss of important prescriptions, and difficulties in navigating to healthcare facilities in unfamiliar environments.

3. **Groups of People Most at Risk**

Traditional Societies and Groups

Indigenous cultures, which frequently live in ecologically fragile places, are especially susceptible to the uprooting that can result from global warming. The deep connection they have to the land, as well as their traditional knowledge and cultural traditions, are in jeopardy as a result of changes in the environment.

Groups with a Low Income and Those on the Margins

Communities with low incomes and populations already on the margins of society are more likely to be affected negatively by climate-induced migration. They frequently lack the means and the capacity for adaptation necessary to deal with the effects of global warming, and they may encounter substantial difficulties in locating housing that is both safe and stable.

Countries Still in the Process of Development

A vast number of poor countries are currently dealing with the repercussions of climate-related migration. The absence of sufficient resources, social safety nets, and infrastructure to sustain displaced populations might make the difficulties they already confront even more difficult to manage.

4. **The Impacts of Climate Change on Human Health**
Illnesses Associated with Heat
Heatwaves are becoming more common and severe as a direct effect of global warming, which has led to an increase in the number of people suffering from heat-related ailments such as heat exhaustion and heatstroke. More at risk are vulnerable populations, such as the elderly and younger generations, especially children.

Diseases Transmitted by Vectors
Temperature shifts as well as shifts in the pattern of the weather have an effect on the behavior and dispersion of disease-carrying insects like mosquitoes. This can result in an increase in the spread of vector-borne diseases such as malaria, dengue fever, and the Zika virus.

The Relationship Between Poor Air Quality and Lung Disease
The effects of global warming on air quality can be made worse by an increase in the amount of ground-level ozone and particle matter. This makes respiratory infections more likely to occur and can make pre-existing problems like asthma and chronic obstructive pulmonary disease (COPD) worse.

Diseases that are Spread by Water
It is possible for water sources to become contaminated as a result of extreme weather events such as flooding, which can then lead to the development of waterborne diseases such as cholera and dysentery. In the aftermath of a natural disaster, gaining access to potable water that is free from contamination is of the utmost importance.

Effects on One's Mental Health
The stress and trauma caused by climate change, displacement, and the loss of homes and livelihoods can result in a variety of

mental health difficulties, including increasing rates of depression and anxiety disorders. One of the main causes of climate change is human activity.

5. Climate-Resistant Methods and Technologies

Adaptation and Protection of Coastal Areas

The implementation of coastal protection measures like sea walls and the restoration of wetland areas can assist limit the amount of displacement that is caused by the rise in sea level. In order for communities and governments to better withstand the effects of extreme weather, climate-resilient infrastructure can be developed.

Preparedness for Emergencies and Emergency Response

The early warning systems and community-based evacuation plans that are part of effective disaster planning and response systems can save lives and reduce the number of people who are forced to relocate. These systems ought to take into account the requirements of vulnerable demographic groups.

Agriculture that is Environmentally Sound

The adoption of sustainable agricultural methods, such as crop types that are resistant to drought and improved water management, can help offset the effects of global warming on food security and on people's ability to make a living.

Healthcare That Is Easily Accessible

It is essential, in order to address the health implications, to ensure access to healthcare services, particularly for populations that have been displaced. Mobile clinics, telehealth services, and outreach programs are all examples of initiatives that can assist close access gaps in healthcare.

Assistance with Mental Health

The psychological effects of global warming and relocation can be mitigated to some extent by the provision of mental health assistance and counseling services to persons who have been relocated. It is important that support systems are easily accessible and take into account different cultures.

The effects of global warming on people are widespread and devastating, causing them to relocate and suffering from a variety of health problems. The most vulnerable groups are frequently the ones that suffer the most devastating effects, often losing their homes, their means of subsistence, and even their lives. Climate change is a worldwide concern that calls for immediate and concerted measures to prevent its causes, adapt to the repercussions of climate change, and provide support to those who are most affected by climate change. A complete response to the human toll of global warming should include the development of communities that are robust to the effects of climate change, the provision of access to healthcare, and the resolution of issues related to mental health. It is vital that societies and governments around the world make these efforts a priority in order to maintain the well-being of persons and communities in every region of the world as average global temperatures continue to rise.

9.2 Economic consequences of climate change and mitigation efforts

The primary cause of climate change is the release of greenhouse gases as a result of human activity. These emissions have significant economic repercussions that are felt all over the world, affecting nations, companies, and individuals. The direct and indirect effects of climate change, such as extreme weather events, rising sea levels, and changes in temperature and precipitation patterns, present difficulties for the global economy. In addition, there are financial repercussions that come with making an effort to slow or stop the progression of climate change by cutting back on emissions of greenhouse gases. This in-depth study investigates the economic repercussions of climate change and the many techniques for mitigating its effects, including the costs of these initiatives and the potential benefits they may offer.

1. **The Impacts of Climate Change on the Global Economy**
   Extremely Harsh Weather Occurrences
   Hurricanes, floods, droughts, and wildfires are all examples of

extreme weather phenomena that have the potential to cause tremendous economic harm. This damage can be caused by the destruction of infrastructure, disruption of supply networks, and financial losses in the regions that are impacted. Damage to property, costs associated with repair and recovery, and insurance claims are all direct outcomes of these occurrences.

The Effects of Agriculture

Agriculture is one industry that may be greatly impacted by climate change, which may lead to decreased agricultural yields, decreased livestock productivity, and increased production costs. Volatility in the price of food has the potential to destabilize economies and contribute to food insecurity, particularly in areas that are highly reliant on agriculture.

Increasingly High Sea Levels

The rise in sea level is a contributing factor in the processes of coastal erosion, floods, and the destruction of important properties along the shore. Property damage, decreased property prices, and greater costs for coastal protection and infrastructure are some of the economic implications.

The Costs of Health

Temperature increases can cause an increase in the number of heat-related illnesses, exacerbate ailments that affect the respiratory system or cardiovascular system, and encourage the transmission of infectious infections. These health impacts lead to increased costs incurred by healthcare providers as well as a decrease in labor productivity.

Vulnerabilities in the Infrastructure

Existing infrastructure, such as roads, bridges, and utilities, may become more susceptible to damage as a result of climate change. This includes both natural and manmade hazards. The cost of repairing and maintaining these facilities can put a strain on public resources and slow the expansion of the economy.

The Process of Moving and Being Moved

The economic hardships that can result from population displacement brought on by the effects of climate change can affect both the displaced population and the host community. It is sometimes necessary to have access to financial resources in order to support and integrate displaced individuals, which can have an impact on local economies and public services.

The Cost of Insurance

It is possible that insurance firms will face increasing liabilities as the frequency and severity of climate-related incidents continues to increase. The ever-increasing cost of insurance premiums can be a burden financially for both people and companies.

2. **Potential Financial Benefits Associated with Climate Change Mitigation**

Renewable Sources of Energy

The shift toward renewable energy sources like solar, wind, and hydropower gives substantial prospects for the economy. Investing in clean energy technology can help economies flourish, and the renewable energy sector has experienced exceptional growth and job creation in recent years.

Effective Use of Energy

Efforts to enhance energy efficiency in companies, buildings, and transportation have the potential to lower energy prices, reduce emissions, and create jobs in industries related to technologies and practices that are energy efficient.

Ecologically Sound Infrastructure

It is possible to encourage economic growth and create new job opportunities through investments in environmentally friendly infrastructure, such as environmentally responsible modes of transportation and green building technology. Projects that use green infrastructure also improve the environment's long-term viability.

Markets for Carbon

The implementation of carbon pricing mechanisms, such as

cap-and-trade systems and carbon taxes, results in the creation of economic incentives for emission reduction. They have the potential to bring in income for the government and encourage enterprises to embrace environmentally friendly technologies and practices.

3. **Obstacles Encountered When Evaluating the Economic Effects**

   Interactions That Are Complicated

   The effects of climate change on the economy are complicated and intertwined with a variety of other issues. Because it is impossible to isolate the economic impact of climate change from other influences, it is difficult to correctly assess the true costs.

   Variation Across Regions

   The economic repercussions of climate change and the efforts that are made to mitigate it might vary substantially from one place to the next. Impacts are typically more severe in vulnerable locations, such as low-lying coastal areas and desert regions, because these types of places are more exposed.

   Lack of assurance

   Because they are dependent on factors such as the rate of global warming, current mitigation efforts, and potential adaptation measures in the future, the long-term economic implications of climate change are susceptible to a large amount of uncertainty.

4. **A Comparison of the Costs and Benefits of Various Mitigation Strategies**

   The Social Impact of Carbon Pricing

   The social cost of carbon (SCC) is an economic metric that is used to estimate the monetary value of damages that arise from each additional ton of carbon dioxide that is emitted into the atmosphere. This metric was developed by the World Resources Institute (WRI) in 2002. It offers a foundation on which to evaluate the advantages of cutting down on emissions.

   Advantages to Avoiding Risks

It is possible for mitigation activities to result in a wide variety of economic benefits, such as decreased healthcare costs as a consequence of an increase in air quality, better energy efficiency, the development of new jobs, and the preservation of ecosystem services that are essential to industries such as agriculture and tourism.

Costs of Taking Precautions

Although there are benefits to be gained from mitigation initiatives, there are also financial expenses associated with them. These include the cost of shifting away from carbon-intensive practices, the cost of establishing carbon pricing schemes, and the cost of investing in clean energy technology.

The Importance of Being Able to Adapt

By making communities more resistant to the adverse effects of climate change, adaptation measures can help mitigate the negative financial effects of the phenomenon. Building infrastructure that is more resistant to the effects of climate change, strengthening public health preparedness, and implementing sustainable agricultural methods are some of these strategies.

5. **The Potential Financial Impacts of Adaptation**
Investment in the Infrastructure

It is generally necessary to make considerable expenditures in infrastructure in order to safeguard coastal areas, enhance water management systems, and ensure robust transportation networks in order to adapt to the effects of climate change. These investments have the potential to boost economic growth as well as the creation of new jobs.

The Cost of Insurance

By lowering the number of climate-related claims filed and their overall severity, proactive adaptation can help bring down the overall cost of insurance. As a result, this can result in additional financial resources being made available for use in other economic endeavors.

Agriculture's Capacity for Resilience

Adaptation in agriculture, such as the adoption of drought-resistant crop varieties and sustainable water management, can maintain food security, stabilize agricultural yields, and promote rural livelihoods. One example of adaptation in agriculture is the adoption of drought-resistant crop types.

6. **International Cooperation and Agreements Concerning the Climate**

Convention de Paris

The Paris Agreement is an international convention that was adopted in 2015 with the goal of reducing the amount of warming that the earth experiences to far below 2 degrees Celsius above its pre-industrial levels. The pact supports global collaboration to address climate change and its economic repercussions by urging countries to cut their emissions and enhance their efforts to adapt to the changing climate.

Markets for Carbon

Trading of emissions allowances across borders is made easier by international carbon markets

such as the European Union Emissions Trading System (EU ETS). These markets not only offer financial incentives for lowering emissions but also have the potential to bring in cash for respective governments.

The effects of climate change on the economy are widespread and widespread, affecting a variety of industries and locations. Some of the most significant difficulties that are posed by global warming include the effects on agriculture, the costs to health care systems, and extreme weather occurrences.

Nevertheless, there are prospects for economic growth in climate change adaptation, particularly in the fields of renewable energy, energy efficiency, and green infrastructure.

Although there are financial costs associated with mitigation initiatives, these costs are typically outweighed by the benefits, which can include cost savings in healthcare, the development of new jobs, and the preservation of the environment. When it comes to addressing climate change and the economic repercussions it has, international collaboration, climate agreements, and carbon markets are all extremely important components. In the end, a holistic strategy for combating climate change, one that takes into account both the reduction of greenhouse gas emissions and the development of strategies to adapt to their effects, is necessary for reducing the economic impact of global warming and developing a more sustainable and resilient future.

9.3 International agreements and cooperation on climate action

Climate change is a global concern that transcends national borders, and in order to address both its causes and its repercussions, it requires concerted efforts and international cooperation from a variety of countries. In order to combat the effects of climate change and encourage participation in collective efforts, a number of international accords and institutions have been developed. This in-depth examination goes into the most important international accords and initiatives on climate action, analyzing their development over time, achievements, and obstacles, as well as the role that global collaboration plays in reducing the negative effects of climate change.

1. **The Immediate Requirement for Collaborative Efforts Across Borders**

   The International Dimension of Climate Change

   The issue of climate change is a global problem that affects every nation, regardless of how much each nation has contributed to the emission of greenhouse gases. Because of the interrelated nature of climate change, a coordinated strategy is required in order to reduce the negative effects of this phenomenon.

   Responsibility that is Shared

   Emerging economies are now important contributors to greenhouse gas emissions, whereas developed countries have historically been the largest contributors to this problem. The idea of "common but differentiated responsibilities" highlights how important it is for both industrialized and developing nations to get involved in the solution to the problem.

Consequences on a Global Scale

The effects of climate change, which include the occurrence of extreme weather, an increase in sea level, and changes to ecosystems, will have repercussions on a worldwide scale. In order to handle these transboundary concerns, international cooperation is absolutely necessary.

2. **The Development of Multilateral Climate Agreements**

The United Nations Framework Convention on Climate Change (UNFCCC) is an international agreement

The United Nations Framework Convention on Climate Change (UNFCCC) was established in 1992 and now serves as the core international framework for addressing climate change. It articulates the overarching principles that should guide climate action, one of which is the objective of maintaining stable amounts of greenhouse gases in the atmosphere.

The Protocol of Kyoto (1997)

As part of the United Nations Framework Convention on Climate Change (UNFCCC), the Kyoto Protocol included legally binding emission reduction objectives for wealthy countries. Despite the fact that it had several shortcomings, it represented a huge step forward in the fight against global warming.

The Agreement Reached in Paris (2015)

The signing of the Paris Agreement in 2015 marked a significant achievement in the field of international climate cooperation. It seeks to keep the increase in global temperature well below 2 degrees Celsius above pre-industrial levels and pursues measures to keep the increase at 1.5 degrees Celsius at most. The agreement does not have any legal weight, but it does lay out a framework for governments to follow in order to establish and meet carbon reduction objectives, encourage adaptation, and improve financial and technological support.

3. **Accomplishments Obtained Through International Climate Agreements**

Commitment on a Global Scale

Countries have been motivated to take action and make pledges in order to reduce their emissions of greenhouse gases as a result of international climate agreements. They have contributed to the development of a sense of communal accountability in the fight against climate change.

Reduced Emissions of Pollutants

The countries that took part in the Kyoto Protocol were able to successfully cut their emissions, which contributed to a decrease in the amount of emissions produced globally. The Paris Agreement expands on these previous endeavors by allowing countries to establish their own goals through the use of "nationally determined contributions," or NDCs.

Help with the finances

Through international accords, institutions for financing climate-related initiatives in developing countries have been formed. These mechanisms aim to facilitate developing nations' transition to low-carbon economies and to support efforts to adapt to climate change.

4. **Obstacles and Restrictions on the Way**

The Administration of Law and Compliance

The absence of enforcement tools to make sure that countries stick to their climate change obligations is one of the main problems with international climate agreements. Because of this, there have been occurrences in which countries have fallen short of their goals for reducing emissions.

Both equity and differentiation are important.

An ongoing difficult issue is the disparities in duties and capacities that exist between countries. In order to combat climate change, developing countries are advocating for increased financial and technological assistance, while industrialized nations are emphasizing the necessity of collective action.

Ambition Discrepancy

Despite the fact that progress has been made, the Nationally Determined Contributions (NDCs) of many nations fall short of what is required to prevent global warming to 1.5 or 2 degrees Celsius. The ambition gap remains a critical challenge that has to be addressed.

Withdrawals and a general lack of involvement

Some nations have made the decision to withdraw from international climate agreements, such as the United States' decision to withdraw from the Paris Agreement during the presidency of Donald Trump. These kinds of measures can make international cooperation more difficult.

5. **The Importance of Collaborating with Other Countries**

Promoting the Development of Innovative Technologies

The creation of new clean energy technologies and innovations can be helped along by international collaboration, which can also speed up these processes. The transition to a low-carbon economy can be accelerated by sharing knowledge and resources with one another.

Increasing One's Capacity

The capacity of developing countries to adapt to and reduce the effects of climate change can

be improved with the assistance of international collaborations. This covers measures relating to the provision of financial support, the transfer of technology, and the strengthening of capacity.

The Trading of Emissions and Offsets

Carbon markets are now possible because to international accords that laid the path for their creation by making it possible for countries and other entities to trade emissions allowances and offsets. These systems provide economic incentives while also encouraging reductions in emissions.

6. **Actors Not Representing the State and Initiatives at the Sub-national Level**

The Part Played by Municipalities and Regions

Even in the lack of leadership at the national level, municipalities and regions all over the world have taken action to curb emissions and adapt to the effects of climate change. They are making a significant contribution to the total decrease in emissions via their efforts.

Participation from the Private Sector

The private sector is absolutely essential to the fight against climate change. Companies are becoming more committed to reducing their carbon footprint by establishing emissions reduction objectives, investing in renewable energy sources, and implementing sustainable business practices.

Civil society and grassroots organizing

Movements at the grassroots level and organizations representing civil society play an important role in holding governments and companies accountable for the climate promises they make. Policies and behaviors relating to climate can be influenced by public pressure.

7. **Agreements and Initiatives of Critical Importance**

The Sustainable Development Goals (or SDGs) of the United Nations

The Sustainable Development Goals (SDGs), and Goal 13 in particular (Climate Action), serve as a worldwide roadmap for addressing climate change and the difficulties that are tied to it. Cooperation on a global scale is required if the SDGs are to be attained.

The Protocol of Montreal

The Montreal Protocol is an effective international agreement that aims to reduce or eliminate the use of compounds that deplete the ozone layer. It is a testament to the efficacy of global cooperation in the face of environmental concerns.

IPCC stands for the Intergovernmental Panel on Climate Change

The Intergovernmental Panel on Climate Change (IPCC) produces credible assessments of climate science in order to assist international climate negotiations. This highlights the significance of scientific collaboration as well as the exchange of knowledge.

8. Moving Ahead with Our Plans

Increasing Our Dedication and Commitment

In order to combat climate change, nations must be incentivized through international climate agreements to set more stringent reduction goals for their emissions. Mechanisms for increased openness and accountability can be helpful in ensuring that nations live up to the commitments they have made.

Providing Assistance to Developing Countries

For the purpose of assisting poor nations in their attempts to adapt to and reduce the effects of climate change, financial support, the transfer of technology, and programs aimed at strengthening capacity should be prioritized.

Converting to an Economy With Less Carbon Dioxide

Countries should enact laws that encourage the use of renewable energy sources, energy efficiency, and sustainable business practices in order to ease the transition to an economy with lower carbon emissions. Innovation in technology and the sharing of knowledge can both benefit from increased international cooperation.

Taking Equity and Differentiation into Consideration

The diversification of obligations and capabilities among countries ought to be a continuing topic of discussion in climate change negotiations. These discussions ought to emphasize the ideas of common but differentiated responsibilities and respective capabilities.

In order to effectively solve the worldwide problem of climate change, international cooperation is of the utmost importance. Even while international climate agreements have achieved great progress, they still have a ways to go before they can be considered successful. These problems and limitations include issues of enforcement, equity, and ambition. When it comes to the fight against climate change, the role that non-state players like cities, regions, the commercial sector, and civil society play is becoming increasingly crucial.

Strengthened international commitments, help for developing countries, and the transition to a low-carbon economy are key components of a sustainable and resilient future in light of the urgent need for the world to limit global warming and reduce the impacts of climate change. The issue of climate change is still one of the most pressing concerns of our time, underscoring the significance of concerted efforts and international collaboration to protect the world for the sake of future generations.

Chapter 10

Mitigation and Adaptation

One of the most major global issues of our day is climate change, which is driven by the growing concentration of greenhouse gases in the atmosphere of the Earth. The effects of climate change, which include an increase in average global temperatures, an increase in the level of the sea, extreme weather events, and disruptions in ecosystems, require a two-pronged strategy for tackling the problem. These strategies are known as adaptation and mitigation. This in-depth investigation delves into the ideas, methods, and essential aspects of both adaptation and mitigation strategies in the battle against climate change. The emphasis is placed on the interdependence of these two types of climate change resistance, as well as the pressing requirement for complete action.

1. **Having an Understanding of Adaptation and Mitigation Adjustment (Mitigation)**

   Actions that lessen or eliminate emissions of greenhouse gases (also known as GHGs) into the atmosphere are referred to as mitigation. Its major objective is to reduce the rate at which the earth is warming and to lessen the severity of climate change's long-term effects. The term "mitigation strategies" refers to a wide variety of actions, including "transitioning to renewable energy sources," "improvement of energy efficiency," and "implementation of policies to reduce emissions."

The process of adapting

The process of adjusting to the effects of climate change in order to minimize vulnerability and build resilience is what we mean when we talk about adaptation. The primary goals of adaptation techniques are to lessen the detrimental consequences of shifting climatic conditions and to assist communities and ecosystems in becoming more resilient in the face of the challenges that climate change presents.

2. **The Dynamics of the Relationship Between Adaptation and Mitigation**

Complementary Methods and Techniques

Both preventing climate change and adapting to its effects are important aspects of climate change management, but they are not mutually exclusive methods. While the goal of mitigation is to address the fundamental causes of climate change, the objective of adaptation is to address the effects of climate change.

Shared Advantages

Numerous solutions for reducing emissions produce co-benefits by simultaneously addressing requirements for adaptation. For instance, making the switch to renewable forms of energy can cut down on emissions and air pollution while simultaneously improving energy security and resiliency.

Conflicting Priorities

In certain circumstances, balancing adaptation and mitigation measures may need making trade-offs in terms of the allocation of resources and the choices that are made. Finding the optimal compromise in climate policy is one of the most significant challenges.

3. **Strategies for Risk Reduction**

Make the Switch to Alternative Energy Sources

A significant component of any mitigation strategy should be the transition away from fossil fuels and toward renewable energy sources such as solar, wind, and hydropower. This results in lower emissions of greenhouse gases and improves energy sustainability.

Effective Use of Energy

Increasing energy efficiency across multiple sectors, including manufacturing, construction, and transportation, results in lower overall energy consumption as well as fewer emissions. In many cases, the implementation of energy efficiency measures results in a speedy return on investment.

The Cost of Carbon

Carbon pricing schemes, such as taxes on carbon or cap-and-trade systems, generate economic incentives for lowering emissions, such as lower energy costs. These regulations bring in income for the government while simultaneously encouraging firms to adopt more environmentally friendly practices.

Both reforestation and afforestation are being done

The removal of carbon dioxide from the air through activities such as reforestation and tree planting is one of the most effective ways to lessen the destructive effects of climate change. The preservation of forested areas is also very crucial.

Transportation That Is Ecologically Sound

Emissions from the transportation sector can be reduced by encouraging the use of public transit, electric vehicles, and cycling in addition to lowering reliance on personal automobiles.

Economy Based on Circulation

A transition to a circular economy, which places an emphasis on recycling and minimizing waste, helps to cut down on the use of resources and the emissions of greenhouse gases that are connected with the manufacturing of new commodities.

4. **Adaptation Methods and Techniques**
   Infrastructure that is Resilient to Climate Change
   Investing in infrastructure that is more resistant to the effects of climate change, such as flood defenses and buildings that can withstand extreme weather, is one way to protect communities and economic assets from the negative effects of climate change.
   Administration of Water Use
   Water scarcity and the increasing danger of floods and droughts are two issues that can be mitigated to some extent by efficient water management practices, which include the creation of water storage and distribution systems.
   Agriculture that is Environmentally Sound
   Food security may be maintained and farming communities' livelihoods can be supported by adopting sustainable agriculture methods such as crop diversity and water-efficient techniques.
   Restoring Balance to an Ecosystem
   Taking measures to maintain and restore ecosystems, such as wetlands, mangroves, and coral reefs, contributes to the conservation of biodiversity and offers natural protection against the effects of climate change.
   Preparedness for Emergencies
   The effects of climate-related disasters on vulnerable communities can be mitigated by the implementation of efficient disaster planning and response systems. These systems include early warning systems and evacuation plans.
   Resilience in Healthcare Systems
   In order to adapt, it is absolutely necessary to improve healthcare systems so that they can deal with the negative effects of climate change on human health. These effects include heatwaves and the spread of diseases carried by vectors.
5. **The Financial Aspect of Climate Change Adaptation and Mitigation**
   The Price Paid for Inaction

If we do nothing about climate change, we will face major economic implications, such as rising healthcare bills, damaged infrastructure, and decreased agricultural productivity.

Possibilities for Profit and Gain

Economic opportunities can be found in both mitigating climate change and adapting to it. The shift toward renewable energy, improvements in energy efficiency, and the adoption of sustainable practices have the potential to generate job growth as well as innovation.

Evaluation of the Costs and Benefits

It is vital, for the purpose of making informed decisions, to conduct an analysis of the costs and benefits of climate action. An examination of costs and benefits can assist governments and businesses in determining the order of priority for investments in mitigation and adaptation.

6. Policies and Administrative Structures

Frameworks on both the National and International Levels

Policies and strategies pertaining to the climate are put into effect by countries using national and international frameworks. The Paris Agreement, which was adopted in 2015, is a global effort to combat climate change. The agreement sets targets for reducing emissions and promotes international collaboration.

Community-Based Initiatives

At the local level, many climate initiatives are taken, with cities and regions serving as primary drivers. These projects frequently place a priority on adaption and mitigation activities that are specifically suited to the requirements and conditions of the local community.

Participation from the Private Sector

The private sector plays an essential part in the fight against climate change. A growing number of companies are embracing environmentally friendly business practices, establishing objectives for the reduction of emissions, and investing in renewable energy and clean technology.

7. Considerations Regarding Ethical Standards And Equity

Climate Change Legitimacy

The concept of climate justice places an emphasis on the moral and ethical aspects of climate change. It asks for an equal distribution of the burdens and benefits of climate action, making certain that vulnerable people are not disproportionately harmed by the actions taken to combat climate change.

A Fair Share of the Burden

Equitable mitigation techniques acknowledge that the responsibility for decreasing emissions should be shared among countries and that developed nations should support developing countries in their efforts. This is in recognition of the fact that developed nations should assist developing nations in their efforts.

Fairness in Adaptation

The requirements of vulnerable groups and countries can be addressed by equitable adaptation methods, which can provide financial support, facilitate the transfer of technology, and organize capacity-building activities.

In the fight against climate change, adaptation and mitigation are two crucial pillars that must be prioritized. In contrast to adaptation, the goal of mitigation is to lessen people's susceptibility to the negative effects of climate change while at the same time limiting the amount of warming that occurs on a global scale. Because of how intertwined and interwoven these methods are with one another, it is necessary to take a comprehensive strategy.

Taking action to combat climate change is not only important from an environmental standpoint; it also has implications for economics, ethics, and government.

A framework for collective action has been established through a number of international agreements and cooperative efforts, including the Paris Agreement and the Sustainable Development Goals.

However, in order to achieve climate targets, not only must the government be committed to doing so, but also the private sector, local efforts, and robust ethical concerns that promote climate justice and equality must be taken into account. To ensure that everyone has a future that is both sustainable and resilient, it is imperative that sustained efforts be made in both mitigating the effects of climate change and adapting to them. The international community is currently confronted with the pressing necessity of limiting global warming and adjusting to its implications.

10.1 Strategies for reducing greenhouse gas emissions

Emissions of greenhouse gases are a major contributor to climate change, which poses a considerable risk to both the natural environment of our planet and the wellbeing of its inhabitants. The international community has come to the realization that there is an immediate and critical requirement to cut these emissions in order to lessen the effects of climate change and make the transition towards a future that is more sustainable and resilient. This essay examines a variety of approaches that might be taken to cut greenhouse gas emissions. Particular emphasis is placed on vital industries such as energy production, transportation, agriculture, and waste management. These tactics include a variety of approaches, such as advancements in technology, regulatory actions, and adjustments to one's way of life.

Transition to Sources of Energy That Are Renewable

The shift away from fossil fuels and toward renewable energy sources is one of the most important steps that can be taken to bring down emissions of greenhouse gases. The principal sources of carbon dioxide emissions, which are the primary contributors to global warming, are fossil fuels. These fossil fuels include coal, oil, and natural gas, amongst other things. It is possible to dramatically cut these emissions by making the switch to renewable energy sources such as solar, wind, hydroelectric, and geothermal power.

Alternatives to fossil fuels such as solar and wind power, in particular, have been making significant strides in terms of cost-competitiveness as well as scalability in recent years. This transformation can be sped up with the help of government incentives and subsidies. The world's nations are setting lofty goals for the widespread adoption of renewable energy sources, and developments in energy storage technology are assisting in making it possible for these sources to provide a supply of energy that is both stable and reliable.

Enhance the Effectiveness of Energy Use

Increasing energy efficiency is another important step that must be taken in order to bring down emissions of greenhouse gases. There is a substantial amount of space for improvement in the energy efficiency of many different types of energy-consuming sectors, including residential, commercial, and industrial. This can be accomplished through the use of strategies such as improved insulation, lighting with LED bulbs, more energy-efficient home equipment, and the use of intelligent technology that reduce overall energy consumption.

In addition, businesses can cut their emissions by improving their operations and minimizing the amount of energy that is wasted. Manufacturing and transportation processes that are more energy efficient have the potential to have a significant influence on the reduction of emissions. Campaigns to raise awareness among the general public, in addition to regulations and incentives that promote energy efficiency, are crucial tools that will drive these advances.

Invest in improving public transportation and raise awareness about the benefits of electric vehicles

The burning of fossil fuels in cars powered by internal combustion engines is the primary cause of the transportation industry's considerable role as a contributor to global warming caused by greenhouse gas emissions. It is vital to stimulate the use of public transportation and to support the deployment of electric vehicles (EVs) in order to achieve a reduction in emissions from this sector.

It is possible to reduce the number of individual automobiles on the road and, as a result, the amount of emissions produced by those vehicles by investing in public transportation systems that are both effective and inexpensive. The shift away from gasoline and diesel vehicles can be sped up with the help of incentives for the adoption of electric vehicles, such as tax credits and the construction of charging infrastructure. It is anticipated that the market share of electric vehicles will increase as technology advances and as the price of EVs continues to drop, which will result in an additional decrease in emissions from transportation.

Put in place a price on carbon

Pricing carbon is a policy method that involves putting a price on greenhouse gas emissions to act as an incentive for people, businesses, and industries to cut back on their overall carbon footprint. There are primarily two ways to put a price on carbon emissions, and those are carbon taxes and cap-and-trade systems.

The imposition of a direct price on emissions, as in the case of carbon taxes, offers a distinct financial incentive to cut down on emissions of greenhouse gases.

On the other hand, systems that use a cap-and-trade approach place a limit on the total emissions allowed and permit businesses to sell emissions allowances. Both of these strategies offer businesses a financial incentive to cut their emissions and increase their investments in technology that are less polluting.

Carbon pricing mechanisms have already been adopted in a number of countries and regions, and as the critical nature of tackling climate change becomes more widely recognized, it is probable that more nations and regions will adopt similar policies.

Agricultural Sustainability Should Be Encouraged

Agriculture is another industry that contributes significantly to greenhouse gas emissions. This is primarily the case because of the methane that is emitted by animals, as well as the carbon dioxide that is generated as a result of changes in land usage and the application of synthetic fertilizers. Agricultural techniques that are more environmentally friendly can contribute to a reduction in the sector's overall emissions.

One of the most important things that can be done is to encourage regenerative agriculture, which puts an emphasis on healthy soil and the capture of carbon. This strategy not only lessens the impact of greenhouse gas emissions but also makes agricultural systems more resistant to the effects of climate change. Reducing the amount of food that is wasted, implementing methods of agroforestry, and moving to organic farming are some other measures that contribute to sustainable agriculture. Organic farming methods typically require fewer synthetic inputs.

Both reforestation and afforestation are being done.

The process of removing carbon dioxide from the air is significantly aided by the presence of forests. Both reforestation, which is the process of replanting forests in regions where they have been cut down, and afforestation, which is the process of establishing new forests in areas where there were previously none, are necessary techniques for the process of sequestering carbon.

These techniques not only contribute to the reduction of emissions of greenhouse gases but also give other advantages to the environment, such as the maintenance of biodiversity and the prevention of soil erosion. Individuals can make a contribution by taking part in tree-planting projects, as can governments and organizations, which can offer financial incentives and other forms of support for reforestation and afforestation initiatives.

Foster Environmentally Responsible Land Use and Urban Planning

Changes in urbanization and land use can have a substantial influence on the emissions of greenhouse gases. Cities are responsible for a major amount of the world's emissions, with variables such as transportation, the energy consumption of buildings, and garbage playing a key influence in the process. It is possible to reduce emissions in urban areas through the use of sustainable land practices and urban design.

It is possible to lessen the need for lengthy commutes and the amount of pollution caused by transportation by encouraging the development of communities that are compact and easily walkable, as well as public transit systems that are effective. In addition, reducing emissions connected with construction and operation of buildings can be accomplished by designing buildings to have a high energy efficiency and promoting the utilization of green spaces. Planning for sustainable urban development contributes to the creation of cities that are environmentally friendly, livable, and have a smaller carbon footprint.

Promote environmentally responsible consumption as well as consumer choices

The actions of consumers have a considerable impact on the production of greenhouse gases. The reduction of emissions is one potential outcome of promoting sustainable consumption and consumer choices. As livestock production is a substantial contributor to methane emissions, one of the strategies that can be implemented is the encouragement of plant-based diets, the reduction of meat consumption, and the support of seasonal and locally sourced food options.

Additionally, encouraging the purchase of energy-efficient appliances and products, as well as reducing trash through recycling and responsible consumption, are all ways to minimize emissions related with the manufacturing, transportation, and disposal of items. Individuals can be empowered to make decisions that are better for the environment by learning more about the environmental impact of their purchasing decisions and the carbon footprints of the things they buy.

Establish Standards for Eco-Friendly Buildings

The erection and continued use of buildings are significant contributors to greenhouse gas emissions. This is essentially the result of the use of energy as well as the utilization of materials that have a high carbon footprint. Green building standards such as LEED (Leadership in Energy and Environmental Design), BREEAM (Building Research Establishment Environmental Assessment Method), and others offer recommendations for the design and construction of structures that are more environmentally friendly and efficient in their use of energy.

Throughout a building's lifecycle, these requirements encourage designs that are more energy efficient, the use of materials that do not deplete natural resources, and a reduction in waste and emissions. In order to cut down on the emissions that are caused by their buildings, businesses and homeowners can voluntarily adopt green building standards, and governments and municipalities can compel compliance with these requirements or offer financial incentives to do so.

Invest in Scientific Exploration and Product Development

It is essential to make investments in research and development (R&D) in order to develop cutting-edge technology and novel solutions for lowering emissions of greenhouse gases. It is possible for governmental bodies, private companies, and academic organizations to all make contributions to this endeavor. Research and development has the potential to lead to significant advances in several fields that are relevant to the reduction of emissions, such as carbon capture and utilization, energy storage, and renewable energy.

Accelerating progress toward a future with lower carbon emissions can be accomplished by providing funding for research in areas such as better battery technology, carbon capture and storage, and sustainable agricultural methods. Additionally, the quick adoption of cutting-edge technology can be facilitated by collaborative efforts involving various levels of government, private industry, and academic institutions.

Taking action to reduce emissions of greenhouse gases is one of the most important problems we face at this moment. This essay presents a comprehensive method to reducing emissions across a wide variety of societal domains through the implementation of the strategies that it outlines. There are several chances to take effective action, ranging from shifting to renewable energy sources and improving energy efficiency to encouraging sustainable agriculture and instituting carbon pricing.

It is absolutely necessary for different levels of government, corporations, communities, and individuals to collaborate in order to put these policies into action and fight climate change as a group. Not only can we lessen the effects of global warming if we cut emissions of greenhouse gases, but we can also build a world that is more robust and sustainable for the children and grandchildren who come after us.

Because of the urgency of the climate catastrophe, we need an approach that takes into account a variety of factors, including advances in technology, changes in public policy, and adjustments in individual conduct. We have the ability to alter the course of the planet's future toward one that is both more sustainable and more prosperous with concerted efforts and a global commitment to decreasing emissions.

The time to take action is now, and the strategies that have been discussed in this article give a road map for accomplishing a major decrease in the emissions of greenhouse gases.

10.2 Approaches to mitigating global warming's effects on ice and seas

The buildup of greenhouse gases in the atmosphere of the Earth causes global warming, which in turn has major consequences on the ice and seas. These effects include the acceleration of sea level rise, the melting of ice sheets and glaciers, and the disruption of marine currents and ecosystems. The effects of these consequences have far-reaching repercussions, not just for the natural world but also for human societies. It will take a strategy with many different components in order to lessen the severity of these effects and protect the future of our planet. This essay explores a variety of tactics and approaches that have been proposed with the goal of reducing the effects that global warming will have on ice and seas.

Cut Down on Emissions of Greenhouse Gases

The decrease of emissions of greenhouse gases should be viewed as the fundamental strategy for minimizing the consequences of global warming on ice and seas. Carbon dioxide (CO_2), methane (CH_4), and other gases that trap heat in the atmosphere are released into the atmosphere when fossil fuels are burned, forests are cleared, and industrial operations are carried out. These gases contribute to the greenhouse effect, which in turn leads to higher temperatures. These higher temperatures, in turn, hasten the melting of ice sheets, glaciers, and permafrost, and cause thermal expansion of seawater.

To find a solution to this problem, governments will need to enact policies and regulations that cut emissions. This involves making the switch to renewable energy sources such as solar, wind, and hydroelectric power; increasing energy efficiency; and promoting the use of electric vehicles (EVs) in order to minimize emissions from transportation. Targets for reducing emissions have been established through international accords such as the Paris Agreement, and nations are obligated to work toward achieving these targets in order to reduce the severity of the impacts of global warming on ice and sea level.

The acronym "Capture and Storage of Carbon" (CCS)

Technologies known as carbon capture and storage (CCS) have the potential to reduce the effects of global warming by sequestering carbon dioxide (CO2) emissions produced by industrial processes and power plants. This would prevent the emissions from being released into the atmosphere. CCS has the potential to play a substantial part in the reduction of emissions and the mitigation of the effects of global warming on ice and seas, in particular in industries where it is difficult to completely eliminate emissions, such as heavy industry.

Both initial forestation and subsequent reforestation

Through a process called photosynthesis, forests are able to remove carbon dioxide (CO2) from the air and act as "carbon sinks." The emission of carbon dioxide into the atmosphere and the alteration of land use are both factors that contribute to global warming. Both afforestation (the process of establishing new forests in places that did not previously have tree cover) and reforestation (the process of replacing forests in areas where they have been chopped down) are essential strategies for storing carbon and reducing the consequences of global warming.

Encourage the Responsible Use of Land and the Development of Coastal Areas

It is absolutely necessary to encourage sustainable land use and coastal development in order

to preserve coastal areas from the threat of increasing sea levels as well as the erosion of shorelines. This involves avoiding construction in regions that are susceptible, adopting building practices that take into consideration rising sea levels and storm surges, and protecting natural protective barriers such as mangrove forests and wetlands.

Put in place the infrastructure for coastal defense

It is vital to invest in coastal defense infrastructure to protect vulnerable communities from the consequences of rising sea levels and increased storm intensity. This can be accomplished by supporting sustainable land use in addition to investing in coastal defense infrastructure. This includes constructing storm surge barriers such as seawalls, levees, and dikes, as well as maintaining them. This type of infrastructure can assist in the prevention of flooding, erosion, and property damage, hence protecting lives and means of subsistence.

Ecosystems along the coasts need to be preserved and restored

The consequences of global warming on ice and oceans can be mitigated, in part, by the coastal ecosystems that include coral reefs, seagrass beds, and salt marshes. These ecosystems provide essential functions. These ecosystems are responsible for the sequestration of carbon, the maintenance of biodiversity, and the performance of a natural buffering function against coastal erosion and storm damage. It is absolutely necessary to take measures to preserve and restore these ecosystems in order to keep their resistance to the effects of climate change intact.

Reduce the Acidification of the Ocean

Ocean acidification is one of the less well-known effects of global warming. It happens when the world's oceans take in more carbon dioxide than they need to and become more acidic as a result. Acidic waters can be harmful to marine life, especially organisms with calcium carbonate shells and skeletons, such as corals, mollusks, and certain types of plankton. Acidic waters can also cause damage to shellfish. Reducing emissions of carbon dioxide (CO_2) and monitoring and regulating the discharge of other pollutants into the ocean are both necessary steps toward mitigating ocean acidification.

Improve the Management of Fish Resources for Sustainable Fishing

Both overfishing and fishing methods that are not sustainable have the potential to wreak havoc on marine ecosystems and make the effects of global warming on seas and ice even worse. The vitality and resiliency of marine ecosystems are supported by practices of sustainable fisheries management. These practices include the creation of marine protected areas, the reduction of unintended catches, and the enforcement of catch limits.

Improve our ability to work together across borders

The effects of global warming on ice and seas are not restricted by national boundaries in any way. Cooperation and coordination on a global scale are very necessary if we are going to be successful in addressing these difficulties. International treaties and organizations, such as the United Nations Framework Convention on Climate Change (UNFCCC) and the Intergovernmental Panel on Climate Change (IPCC), play critical roles in facilitating collaboration between nations, sharing information and best practices, and establishing global climate objectives.

Adaptational Steps to Take

The ongoing and unavoidable consequences of global warming on ice and seas require, in addition to actions aimed at mitigating those effects, the implementation of measures aimed at adapting to those effects. These methods assist communities and ecosystems in adjusting to the shifting conditions and safeguard groups that are particularly susceptible.

Infrastructure that is resistant to the effects of climate change: Coastal towns can be protected against floods and erosion by investing in infrastructure that can withstand the effects of increasing sea levels. Examples of such infrastructure include flood barriers, higher structures, and enhanced drainage systems.

Establishing early warning systems for extreme weather events and rising sea levels can help communities prepare for and respond to hazards, therefore lowering the potential for loss of life and property. These systems can help communities prepare for and respond to threats.

Sustainable agriculture and water management: Adapting agricultural techniques and water management systems to adapt to changing climate conditions can assist assure food security and lessen demand on water resources. Both of these goals contribute to sustainability in agriculture and water management.

Conservation of biodiversity: Maintaining the resiliency of coastal areas and helping to preserve biodiversity at the same time by safeguarding and rehabilitating ecosystems such as mangrove forests, wetlands, and coral reefs.

Human displacement and community engagement: In certain instances, it may be required to relocate communities that are at a high risk of being affected by both rising sea levels and erosion of the coastline. It is possible to make the process more egalitarian and successful by including communities that are affected in the decision-making process and offering assistance for relocation.

The effects of global warming on ice and seas are one of the most important concerns facing our generation at this moment. There will be far-reaching repercussions for both the environment and for human societies as a result of melting ice sheets, increasing sea levels, ocean acidification, and the disturbance of marine ecosystems. In order to lessen the severity of these effects, a multifaceted strategy is required. This strategy must include the reduction of emissions of greenhouse gases, the implementation of adaption measures, and the strengthening of international collaboration.

We can limit the rate of ice melt, mitigate sea-level rise, and reduce the impacts on seas if we address the underlying causes of global warming and take efforts to protect vulnerable regions, ecosystems, and populations. In other words, we can address the core causes of global warming. In the end, the future of the ice and seas on our planet is contingent on our ability to jointly respond to the challenges posed by global warming and to conserve these essential components of our global ecology.

10.3 Efforts to adapt to a changing climate, both locally and globally

The effects of climate change are one of the most urgent problems that the world is currently experiencing. The rate at which the climate of the Earth is shifting is unprecedented, and it is mostly attributable to human activities such as the combustion of fossil fuels, the clearing of forests, and various industrial processes. Because of the widespread effects these changes will have on ecosystems, economies, and societies, it will be necessary to respond in a way that is both coordinated and comprehensive. Adaptation is a process of adjusting to new and changing climate circumstances in order to limit the harm and maximize the potential opportunities that may arise as a result of these changes. Adaptation is an essential component of the fight against climate change.

At both the local and the global level, it is imperative that efforts be made to adapt to the changing climate.

In the following paragraphs, we will discuss a variety of approaches, programs, and plans that are being implemented at the individual, community, and national levels in order to adapt to the effects of climate change. These initiatives involve a wide variety of techniques, some of which include the construction of climate-resilient infrastructure, the protection of ecosystems, the improvement of agricultural practices, and the enhancement of disaster preparedness.

Efforts Made at the Community Level to Adapt to Climate Change

Infrastructure that is Resilient to Climate Change

Building and improving climate-resilient infrastructure is an important component of any local adaptation strategy, particularly in light of the fact that climate change is expected to bring about increasingly frequent and severe weather events. This involves the construction of flood barriers, the strengthening of buildings to resist more intense storms, and the improvement of drainage systems to lessen the likelihood of flooding. In order to defend themselves against both rising sea levels and storm surges, a number of coastal cities are currently engaged in the process of raising vital infrastructure and building seawalls. Investments in infrastructure of this kind are absolutely necessary in order to protect communities from the direct effects of climate change.

Systematic Methods of Early Warning

A lot of local communities are working on creating early warning systems in order to get ready for the extreme weather events and natural disasters that are expected to occur as a result of climate change. In order to offer people and authorities with timely warnings, these systems make use of sophisticated meteorological data and communication technologies. Early warning systems assist in the evacuation of vulnerable populations, which in turn helps to save lives, decrease property damage, and prevent further injury. Increasing local resilience to the effects of climate change requires the kind of initiatives that are being discussed here.

Planning for the Sustainable Use of Land and Urban Space

One other essential strategy for local climate adaptation is to prioritize the implementation of sustainable land use and urban planning practices. In order to accomplish this, green spaces need to be created, wetlands and natural barriers need to be preserved, and building methods need to be adopted that take into account the impact of climate change. The urban heat island effect is one of the primary targets of sustainable urban design, along with the reduction of flooding and an overall improvement in the citizens' quality of life.

Administration of Water Use

The variations in precipitation patterns that are commonly caused by climate change might result in rain that is both more intense and more variable. The implementation of efficient water management methods, such as the collection of rainfall through the building of rainwater harvesting systems and improved control of stormwater runoff, can help reduce the risk of flooding and guarantee a consistent water supply even during times of drought. In addition, actions taken to enhance water quality and minimize pollution are beneficial to the overall resilience of communities.

Efforts Made to Preserve and Rehabilitate Ecosystems

The effects of climate change can be partially mitigated by natural ecosystems like mangrove forests, wetlands, and coral reefs, which all play an important part in the process. These ecosystems not only serve as natural barriers to protect the coast from erosion, but they also provide habitats for a wide variety of animal species. The biodiversity of these ecosystems is increased as a direct result of local efforts to protect and restore them, which also helps communities become more resilient to the effects of things like rising sea levels and extreme weather.

Agriculture that is Environmentally Sound

Agriculture is one sector that may be adversely affected by climate change, which may result in different rainfall patterns, higher temperatures, and an increase in the number of pests and illnesses. The promotion of drought-resistant crop varieties, the installation of water-efficient irrigation systems, and the development of agroforestry practices that promote soil health and carbon sequestration are all examples of local adaptation initiatives in the agricultural sector. In the face of climate change, communities that practice sustainable agriculture have a better chance of protecting their food supply.

Human Relocation and Participation in Community Activities

The effects of climate change may, in certain instances, require communities to seriously contemplate moving elsewhere. Moving to higher ground might be the only practical choice for people living in low-lying locations that are vulnerable to rising sea levels and erosion of the coastline. However, because they include the uprooting of communities as well as the activities that are integral to those communities, such relocations are complicated and call for meticulous planning. It is possible to make the process more egalitarian and successful by including communities that are affected in the decision-making process and offering assistance for relocation.

Preparedness for Emergencies and Emergency Response

In light of climate change, more and more communities are coming to the realization that it is very important to have solid programs for disaster planning and response. This involves activities such as the training of first responders, the development of evacuation plans, and the storage of emergency supplies. In addition, governments are working to increase the resiliency of their communities by participating in public awareness initiatives and teaching locals on how to best prepare for and respond to the effects of extreme weather.

Efforts Made All Over the World to Accommodate Climate Change

Convention of the United Nations Framework on Change in the Atmosphere (UNFCCC)

The United Nations Framework Convention on Climate Change is an important worldwide framework for dealing with climate change. Its sessions, which are known as the Conference of the Parties (COP), bring together different governments to discuss and negotiate climate action. The United Nations Framework Convention on Climate Change (UNFCCC) ratified the Paris Agreement in 2015 with the intention of reducing the rate of global warming to far below 2 degrees Celsius over pre-industrial levels. The agreement outlines international obligations to limit emissions of greenhouse gases and support efforts to adapt to climate change, particularly in developing nations that are particularly susceptible.

IPCC stands for the Intergovernmental Panel on Climate Change

The Intergovernmental Panel on Climate Change (IPCC) is a scientific organization that has been entrusted with evaluating the most recent findings in the field of climate science and providing policymakers with information on the potential implications of climate change. It generates reports that contribute to international climate negotiations and help guide plans for climate adaptation and mitigation.

Fund for the Green Climate (GCF)

The Global Climate Fund was formed to assist developing countries in adapting to the effects of climate change and minimizing the negative impacts of climate change. It does this by providing funds for projects and programs that attempt to increase the resilience of ecosystems and communities that are particularly vulnerable. The Global Climate Fund (GCF) works to enlist the financial assistance of wealthy nations in order to assist less developed states in meeting the problems posed by climate change.

Fund for Adaptation

The Adaptation Fund is another financial vehicle that was formed in order to fund climate adaptation projects in developing countries. It was established as part of the Kyoto Protocol.

It does this by awarding grants and granting preferential interest rates on loans to initiatives that are designed to make communities and ecosystems more resilient to the effects of climate change.

Plans for National Adaptation (also known as NAPs)

A significant number of countries are in the process of formulating their National Adaptation Plans (NAPs), which describe how they intend to adjust to the effects of climate change. NAPs assist countries in assessing their own vulnerabilities, determining which adaptation priorities are most important, and allocating resources to put these priorities into action. At the national level, the provision of a systematic framework for climate adaptation is made possible, in large part, by the creation of these plans.

Insurance Against Climate Risk

Insurance against the risks posed by climate change is a forward-thinking worldwide project that helps economically vulnerable communities prepare for and recover from climate-related calamities. It provides a safety net for people who are experiencing the negative effects of climate change and helps them recover from the financial losses that are associated with extreme weather events.

Transfer of Technologies and Enhancement of Capabilities

The transfer of technologies to poor countries and the strengthening of their capacity to adapt to climate change are both part of the global effort to adapt to the effects of climate change. This includes providing these nations with assistance in the form of knowledge, skills, and technology in order to assist them in building the capacity to effectively implement adaptation methods.

Fostering International Cooperative Relationships and Solidarity

The ability to adapt to a changing global environment requires significant amounts of international collaboration and solidarity. Developed countries, which are responsible for a substantial share of the world's historical emissions of greenhouse gases, have a moral duty to assist less developed countries in their efforts to adapt to climate change. The assistance that industrialized countries can provide to vulnerable regions might come in the form of financing, the transfer of technology, and the strengthening of capability.

Both difficulties and prospects are involved

Adaptation efforts to a changing climate, whether on a local or global scale, confront a number of problems and opportunities of varying degrees.

The Obstacles:

Limited resources: Many communities, particularly in developing nations, lack the financial resources and infrastructure necessary to effectively adapt to the effects of climate change. This is especially true in the United States. There have been global efforts made, such as the Green Climate Fund, with the intention of addressing this problem; nevertheless, additional funds and assistance are required.

Concerns concerning fairness and justice have been raised as a result of the disproportionate impact that climate change has on populations who are already vulnerable or disenfranchised. It is of the utmost importance to make certain that adaptation efforts are equitable and do not worsen inequalities that already exist.

Ambiguity: Because climate change is characterized by high ambiguity, it is challenging to forecast the precise type and timing of its repercussions. This makes it difficult to foresee the future. The planning and adaption processes may become more difficult as a result of this uncertainty.

Interests at odds: In the context of international discussions, differences in interests and objectives among nations can be a barrier to making forward in attempts to adapt to a changing environment. The complicated and constant problem of negotiating climate agreements that fulfill the needs and concerns of all nations is one that must be tackled.

Occasions to seize:

Innovation: Finding solutions to climate change opens the door to new possibilities in both technology and policy innovation. There is a constant emergence of novel techniques and solutions that have the potential to boost the efficiency of current adaptation efforts.

Co-benefits: In addition to mitigating the negative effects of climate change, many climate adaptation measures also confer a number of additional advantages. Reforestation and other sustainable land management strategies, for instance, have the potential to increase biodiversity, improve soil health, and sequester carbon.

Sharing of knowledge It is essential for effective climate adaptation that communities, nations, and international organizations share their knowledge with one another as well as discuss and implement best practices. It is possible to expedite the adoption of effective tactics by gaining knowledge from the experiences of others.

Awareness and advocacy: As people become more aware of the impacts of climate change, there is an increase in support for attempts to adapt to these changes. The mobilization of resources and support for these projects can be facilitated by advocacy and awareness-raising campaigns.

The pressing issue of climate change is a worldwide concern that calls for coordinated actions on a local and global scale to adapt to its effects. A multidimensional strategy is required, beginning at the local level with the development of climate-resilient infrastructure and early warning systems and continuing all the way up the hierarchy to the global level with international agreements, funds, and projects. The ability of people, ecosystems, and economies to adapt to climate change is absolutely necessary in order to defend themselves from the negative effects of a changing climate. Additionally, this presents a chance to advance sustainability, fairness, and innovation while simultaneously constructing a world that is more climate-ready and resilient. As the effects of climate change become more apparent, it is more important than ever for individuals, communities, governments, and the international community to collaborate in order to adjust to the newly established climate reality and ensure a sustainable future for everyone.

Chapter 11

Hope and Solutions

It is crucial to acknowledge that hope is not just a powerful emotion but also a catalyst for change in a society that is confronted by global challenges that are complicated and linked. This is especially important given the state of the world today. Hope offers the motivation, resilience, and inspiration necessary to drive progress toward a goal. This is true whether the goal is to combat climate change, eliminate poverty, or address issues in public health. This essay goes deeply into the complex problems that we are confronted with on a worldwide scale. It then investigates the potential answers to these problems and places an emphasis on the role that hope plays in finding a solution.

1. The Changing Climate

There is little doubt that climate change is one of the most pressing global concerns. Its effects on the environment, cultures, and economy are all expected to have far-reaching repercussions. A global call to action has been prompted by the many repercussions of climate change, including rising temperatures, an increase in the level of the sea, extreme weather events, and the loss of biodiversity. Nevertheless, combating climate change needs more than simply acknowledging the issues at hand; rather, it calls for the development of actionable solutions as well as the promotion of an optimistic vision for the future.

1. **Strategies for Risk Prevention**
   Make the Switch to Alternative Energy Sources
   The transition away from fossil fuels and toward renewable energy sources like solar, wind, hydroelectric, and geothermal power is one of the most important things that can be done to slow or stop climate change. The swift growth of clean energy infrastructure and technology, together with government incentives and supportive regulations, provides reason for optimism over the possibility of a future free of carbon emissions.
   The Cost of Carbon
   Carbon pricing mechanisms, such as carbon taxes and cap-and-trade systems, provide economic incentives for individuals, enterprises, and industries to reduce their carbon footprints. These economic incentives are provided by carbon pricing mechanisms. These programs give hope for a successful plan to combat climate change by helping to limit emissions and funding additional climate action.
   Both reforestation and afforestation are being done
   Forests serve as carbon sinks because of their ability to absorb and store carbon dioxide. Both reforestation (the process of restoring trees in areas where they have been removed) and afforestation (the process of developing new forests) are viable solutions for capturing carbon and battling climate change. The only way for the planet to become healthier is if global efforts are made to preserve and restore these important ecosystems.
2. **Adaptive Methods and Techniques**

Infrastructure that is Resilient to Climate Change

The construction of infrastructure that is resistant to the effects of climate change, such as flood barriers, raised buildings, and enhanced drainage systems, assists communities in withstanding the effects of climate change. The knowledge that putting money into such infrastructure will protect vulnerable communities and cut down on the harm caused by natural disasters gives people hope.

Systematic Methods of Early Warning

Communities are better able to plan for and respond to severe weather disasters when they have access to early warning systems that make use of advanced meteorological data and communication technologies. These systems play a critical role in preventing loss of life and property and enhancing a community's resistance to the effects of climate change.

Agriculture that is sustainable, as well as water management

In order to maintain food security in the face of climate change, it is vital to adapt agricultural methods. Some examples of such adaptations include drought-resistant crops and water-efficient irrigation systems. Furthermore, sustainable water management practices can prevent flooding and safeguard water supplies during dry years, presenting hope for an agricultural sector that is more resilient.

II. Economic Disparities and Deprivation

Poverty and inequality are pervasive issues that need to be addressed using a variety of different approaches. Initiatives aimed at eliminating poverty and narrowing the gap between rich and poor offer some glimmer of hope for a society that is more equal.

1. **The Promotion of Economic Growth**
   Education and Training Leading to Professional Growth
   People are given the tools they need to climb out of poverty and take part in the global economy when resources are invested in education and programs that build their skills. To give everyone the opportunity to receive a good education is the foundation on which to build optimism for a better future.

Assistance with Micro Financing and Small Businesses

The provision of access to cash and financial services for underserved communities by means of microfinance and other small business support programs gives individuals the ability to improve their economic prospects and gives them greater agency. These programs inspire entrepreneurship and a sense of independence, which in turn gives people reason to have optimism.

2. **Social Protection and Safety Nets**

UBI stands for "universal basic income"

The term "Universal Basic Income," or UBI for short, refers to the idea that every citizen should be entitled to a predetermined amount of money from the government on a consistent basis. The universal basic income (UBI) provides a glimmer of hope by leveling the economic playing field and eliminating income disparity.

Cost-Effective Medical Care

It is a fundamental right, as well as a crucial component of efforts to reduce poverty, to have reasonable access to medical care. The best hope for enhancing existing healthcare systems is the implementation of universal healthcare systems and cost-effective medical services.

3. **Equality of the Sexes**

In order to reduce inequality on a worldwide scale, it is vital to promote gender equality. There is reason to have hope for a more diverse and inclusive world if we promote women's empowerment, fair pay, and representation in positions of leadership.

III. The Health of the Public

There has never been a time when the problems facing public health, such as pandemics, access to treatment, and emerging illnesses, have been so obvious.

In spite of this, the combination of collaborative efforts and inventiveness might offer a glimmer of optimism in the face of these difficulties.

1. **The Preparation for a Pandemic**
   Collaborations on a Global Scale Regarding Health
   Hope is provided by international organizations and alliances like the World Health Organization (WHO) and projects like COVAX, which coordinate the response to global health emergencies, provide access to vaccines, and share crucial information with one another.
   The Activities of Research and Development
   Spending money on research and development is absolutely necessary in order to combat newly discovered diseases. There is reason for optimism in the fight against dangers to public health thanks to developments in diagnostics, treatments, and immunizations.
2. **Healthcare for All of the People**
   The resilience of the public health system is dependent on the provision of healthcare for all. It is possible for all people, regardless of their socioeconomic standing, to have access to medical services if we have healthcare systems that are both accessible and inexpensive. This gives people hope.
3. **Education Concerning Health**

Individuals and communities gain the ability to make well-informed decisions that are protective of their wellbeing when health education, preventative, and immunization initiatives are actively promoted. The only glimmer of hope rests in programs that educate people and encourage them to adopt healthier habits.

IV. Protection of the Natural Environment

The preservation of the natural world is absolutely necessary in order to guarantee a prosperous future. Efforts to preserve the environment and creative problem-solving offer a glimmer of hope for a world that will continue to support life for future generations.

1. **The Protection of Flora and Fauna**
    Reserves and Other Forms of Conservation Land
    The creation of protected areas and wildlife reserves is an encouraging step toward preserving biodiversity and preventing the extinction of threatened species.
    Methods That Are Ecologically Sound
    It is possible to lessen the detrimental effects that human activities have on ecosystems by promoting environmentally friendly agricultural, forestry, and fishing methods. These methods provide reason for optimism regarding the protection of natural resources.

2. **Circular Economy**
    A more sustainable and less resource-intensive future may be attainable through the implementation of a circular economy, which calls for the reduction of waste while also emphasizing the reuse and recycling of available resources.

3. **Innovation and technological advancement**

There is reason to be optimistic about a future in which human activities and natural systems may coexist without inflicting any damage as a result of technological advancements in the fields of clean energy, waste management, and resource-efficient technology.

V. Conflicts on a Global Scale

There is always hope for peace and reconciliation, despite the fact that violent conflict and unstable states are huge worldwide issues.

1. **Diplomacy and Other Forms of Conciliation**
    There is reason to have hope that conflicts may be resolved, collaboration can be fostered, and peaceful resolutions can be

reached through diplomatic efforts and international mediation mechanisms.

2. **Conflict Avoidance and Resolution**
   There is reason to have hope for lessening the chance of conflicts breaking out as a result of investing in conflict prevention techniques such as efforts aimed at establishing peace and programs designed to alleviate poverty.

3. **Aid to People in Need**

The alleviation of pain and the provision of critical support to populations that are vulnerable is what gives people hope when humanitarian help is provided to individuals who have been affected by conflicts.

VI. Technological Advancement and Creative Effort

The development of new technologies and the introduction of novel ideas hold the promise of

resolving a plethora of pressing problems on a global scale and giving rise to optimism over a brighter future.

1. **Technologies that don't hurt the environment**
   There is reason to be hopeful about mitigating the negative effects of climate change, thanks to recent developments in environmentally friendly technology such as green energy, carbon capture, and environmentally responsible transportation.

2. **Breakthroughs in Health Care Innovations**
   There is reason to be hopeful about improvements in both public health and medical care brought about by recent developments in healthcare technologies, diagnostics, and treatments.

3. **Information and Communication Technology**

Education could be improved, the digital divide could be bridged, and global connection could be fostered all with the use of information technology and increased digital connectivity.

VII. Education and Sense of Responsibility

Education and initiatives to raise awareness are vital components of any strategy to address global concerns. They inspire individuals and communities to take activities that are informed by relevant information and give hope for positive change.

1. **Education Regarding the Climate**
   Education about climate change offers a glimmer of hope since it raises people's knowledge of environmental problems, encourages environmentally responsible behaviors, and helps to raise a new generation of environmental stewards.
2. **The Promotion of Social Justice**
   The promotion of social justice and equality acts as a catalyst for change, presenting the prospect of a future that is more equitable and welcoming to all.
3. **Literacy in the Scientific Domain**

Individuals will have the ability to interact with complex global concerns, which will inspire hope through informed decision-making if scientific literacy and critical thinking are encouraged and promoted.

VIII. Cooperative Efforts on a Global Scale

The ability for states to work together for the common good and the ability to inspire hope for a more peaceful and sustainable world are both enabled by international cooperation and diplomacy, which are key components of the process of addressing global concerns.

1. **Agreements Reached Between Countries**
   The Paris Agreement, the Sustainable Development Goals (SDGs), and the International Health Regulations are examples of multilateral agreements that provide a glimmer of hope since they establish international norms and objectives for tackling global concerns.
2. **The Organization of the United Nations and Other International Bodies**

The United states and other international organizations provide people reason to have optimism because they help states communicate with one another, negotiate agreements, and work together to find solutions to global problems.

Hope is not simply a feeling but a driving force for positive change in a world that is defined by complicated and interconnected global concerns. It strengthens our will to combat climate change, end global poverty, defeat public health problems, protect the environment, find peaceful solutions to conflicts, and welcome advances in technology. Hope has the power to change the world by igniting solutions to the problems that face the entire planet. This change can come about through education, innovation, international cooperation, and individual action. As we move forward, let us take heart from the power of hope and collaborate with one another to create a future that is more sustainable, equitable, and peaceful for everyone.

11.1 Success stories of communities and nations taking action

In an era that has been distinguished by difficulties and complexities that have never been seen before, the globe has witnessed an extraordinary number of success stories involving communities and nations working together to take preventative action. The strength of collective action has shown that it has the potential to transform situations, whether it be in the fight against poverty, the promotion of sustainable practices, or the resolution of social inequities. These examples of success serve as rays of light and sources of motivation by demonstrating the measurable impact that can be accomplished via the coordinated efforts of multiple parties and the implementation of strategic initiatives.

After the horrific events of the genocide that occurred in 1994, the society of Rwanda underwent a tremendous metamorphosis, which is one of the most astonishing success stories. Rwanda has made great progress toward both economic development and reconciliation, despite the fact that the country has been ripped apart by ethnic warfare. The nation has been able to successfully restore its social fabric and establish itself as a symbol of resilience as a result of a mix of strong leadership, community participation, and international help. The focus placed by the government on developing a culture of unity and inclusivity has played a vital role in fostering a sense of collective identity among the Rwandan people, opening the way for prolonged peace and progress. This has been one of the most important factors in the country's recent history.

In a similar vein, the countries of Scandinavia, such as Sweden, Norway, and Denmark, have emerged as global role models in terms of the progressive social policies they have enacted and the commitment they have made to sustainable development. These countries have achieved extraordinary success in promoting social equality and environmental sustainability thanks to the combination of strong welfare systems, good governance, and a strong emphasis on environmental conservation. Not only has their proactive approach to addressing issues such as healthcare, education, and gender equality enhanced the quality of life for their residents, but it has also acted as a pattern for other countries who are attempting to develop societies that are more egalitarian and sustainable.

In addition, the success story of the Bhutanese model of Gross National Happiness (GNH) provides as a striking example of a nation placing a priority on the well-being of its population over the expansion of its economic base. Bhutan has been successful in striking a delicate balance between the country's rapid modernization and the protection of its distinct cultural legacy because of the country's emphasis on the significance of holistic development and sustainable living. The commitment of the Bhutanese government to fostering happiness as a vital indication of success has drawn international attention and admiration, underscoring the value of prioritizing the total well-being of society over material prosperity. This devotion has garnered international attention and appreciation.

The success story of Costa Rica in the area of environmental conservation stands out as a monument to the transforming potential of sustainable policies and conservation initiatives. This is demonstrated by the country's status as an environmental leader. As a result of proactive steps taken to preserve its rich biodiversity and natural resources, Costa Rica has emerged as a world leader in environmentally responsible ecotourism and the protection of the natural environment. It is thanks to the ambitious reforestation programs undertaken by the country as well as its dedication to the use of renewable energy sources that the nation has been able to considerably cut its carbon footprint and become a model for environmental stewardship in the eyes of the rest of the globe.

In addition to this, the world community's reaction to the COVID-19 epidemic brought to light the importance of international cooperation and concerted effort in the fight against global catastrophes. In spite of the enormous difficulties faced by the epidemic, a large number of communities and nations joined forces to work together on the production of vaccines, their distribution, and ensuring that everyone had equal access to them. The speedy implementation of vaccination campaigns, along with the concerted efforts to lessen the socioeconomic burden of the epidemic, brought to light the significance of solidarity and unity in facing common challenges.

Furthermore, the success stories of grassroots movements and community-based projects all over the world have proved the transformative power of local action in solving social challenges and fostering good change. These tales have been collected in a book titled "The Success Stories of Grassroots Movements and Community-Based Initiatives." These efforts have demonstrated the ability of collective agency to drive significant change at the grassroots level, and they range from community-led education programs in rural areas to grassroots environmental conservation projects. These projects have contributed to the empowerment of marginalized groups and the promotion of sustainable development on a local scale. This has been accomplished via the mobilization of resources, the promotion of collaboration, and the empowerment of local communities.

11.2 Innovations in renewable energy and climate-friendly technology

Solar Power Industry Developments:

In the past few years, tremendous technological advancements have been made in the realm of solar energy, which is one of the most abundant and easily available forms of renewable energy. The development of solar panels with high conversion efficiencies, such as monocrystalline and polycrystalline silicon cells, has led to a huge increase in the amount of sunshine that can be converted into power. In addition, the development of thin-film solar cells and organic photovoltaic technologies has opened up new opportunities for integrating solar panels into a variety of surfaces, such as the facades of buildings, windows, and even clothes. This has increased the potential for mass acceptance of solar energy and broadened the scope for its use.

In addition, the development of solar tracking systems and concentrated solar power (CSP)

technologies has made it possible to utilize solar energy in a more effective manner. These technologies increase the amount of sunlight that is captured and use that sunshine to generate electricity. In addition, the incorporation of artificial intelligence (AI) and data analytics into solar energy management has resulted in an improvement in the operational efficiency of solar power plants, which has led to optimal energy output and greater grid stability.

Recent Developments in Wind Energy:

In the field of wind energy, developments in turbine technology and offshore wind farms have revolutionized the possibility for harvesting wind power on a massive scale. This potential was previously limited by the physical limitations of the wind. The production of energy by wind farms has substantially improved, as has their level of dependability, thanks in large part to the invention of wind turbines that are larger, more powerful, and better equipped with cutting-edge blade designs and control systems. Furthermore, the construction of floating wind turbines in deep waters has increased the geographical scope for the generation of wind energy, enabling the tapping of wind resources in places that were previously judged unsuitable for traditional fixed-bottom offshore wind installations. This has resulted in an expansion of the geographical scope for the generation of wind energy.

In addition, advancements in wind energy storage devices, such as grid-scale batteries and pumped hydro storage, have helped mitigate the intermittent nature of wind power, which has made it possible for wind energy to be more effectively incorporated into preexisting power networks. Integration of smart grid technology and predictive maintenance systems has significantly improved the overall efficiency and dependability of wind energy infrastructure, making it possible for a smoother transition towards a power supply that is reliant on renewable energy sources.

Recent Advances in Bioenergy:

Significant progress has been made in the production of biofuels and biogas, two areas that are important to the whole field of bioenergy, which is generated from organic resources such as agricultural residues, organic waste, and crops specifically grown for energy generation. The manufacture of cleaner and more environmentally friendly alternatives to conventional fossil fuels has been made possible by the development of cutting-edge technology for the conversion of biofuels, such as biochemical and thermochemical processes. Inventions in the field of biofuels, such as cellulosic ethanol and biofuels derived from algae, have demonstrated significant potential for lowering emissions of greenhouse gases and minimizing the negative effects on the environment that are caused by the use of traditional fuels.

In addition, the adoption of anaerobic digestion and biogas purification technologies has made it easier to convert organic waste effectively into biogas, a sustainable energy source that can be utilized for the generation of both heat and power. Integrated biorefinery systems, which optimize the utilization of various feedstocks for the production of bioenergy and bioproducts, have contributed to the advancement of a circular bioeconomy. This promotes resource efficiency and sustainability across a variety of sectors, including agriculture, forestry, and waste management, among others.

Innovations in the Storing of Energy:

The development of improved energy storage technologies has emerged as a vital enabler for the broad adoption of renewable energy sources. These technologies aim to solve the issues given by the intermittent nature of renewable energy sources as well as the instability of the grid. The efficiency, energy density, and lifespan of energy storage solutions have all seen significant improvements as a result of advancements in battery storage systems. These advancements, which include lithium-ion batteries, flow batteries, and solid-state batteries, have made it possible for variable renewable energy resources to be seamlessly integrated into the power grid.

Additionally, the development of hydrogen storage and fuel cell technologies has prepared the way for the exploitation of hydrogen as a versatile and clean energy carrier. This facilitates the decarbonization of different sectors, including the transportation industry, industrial production, and power generation. The discovery of methods for the generation of hydrogen that are both scalable and cost-effective, such as electrolysis powered by renewable energy sources, has reinforced the possibilities of an economy based on hydrogen. This offers a possible pathway towards the achievement of long-term sustainability and energy independence.

The Internet of Things with Intelligent Energy Management Systems:

The optimization of the efficiency and dependability of the infrastructure supporting renewable energy sources has been significantly aided by the incorporation of technology for smart grids and advanced energy management systems. In the context of variable renewable energy generation, the adoption of intelligent grid control systems, which include real-time monitoring, predictive analytics, and demand response mechanisms, has made it possible to improve grid management and strengthened the stability of power supply. This is especially true in light of the fact that renewable energy generation can be highly variable.

In addition, the implementation of Internet of Things (IoT) devices and sensor networks in energy systems has made it easier to collect real-time data and to put predictive maintenance strategies into place. These developments have led to an improvement in the operational performance of renewable energy assets and a reduction in the amount of downtime those assets experience. The use of blockchain technology in energy trading and peer-to-peer energy transactions has also fostered the development of decentralized energy systems. This has enabled consumers to actively participate in the production and distribution of renewable energy, thereby promoting energy democratization and sustainability. Blockchain technology was initially developed to record and verify transactions in digital currencies.

Transportation electrification with environmentally responsible practices:

The electrification of vehicles and the development of sustainable mobility solutions have acquired substantial pace in the transportation sector, which has catalyzed the transition toward cleaner and more efficient transportation systems. The development of electric vehicle (EV) technologies, such as high-capacity batteries, fast-charging infrastructure, and regenerative braking systems, has accelerated the adoption of electric cars, buses, and commercial vehicles. This has contributed to the reduction of greenhouse gas emissions and air pollution in urban areas. Electric vehicles have also contributed to the acceleration of the adoption of electric cars.

Furthermore, the development of autonomous and connected vehicle technologies has revolutionized the concept of smart mobility. This has fostered the integration of electric and shared transportation services, such as electric scooters, bikes, and ridesharing platforms. This has led to the promotion of sustainable urban mobility and the reduction of the environmental footprint caused by transportation activities. The development of sustainable transportation ecosystems has been made easier by the convergence of infrastructure for the charging of electric vehicles and the generation of renewable energy. This has resulted in increased energy efficiency, protection of natural resources, and resistance to the effects of climate change in the global transportation network.

Policy Considerations and Market Motives:

In addition to developments in technology, the establishment of encouraging regulatory frameworks and the introduction of market incentives have been critical factors in pushing the acceptance and deployment of climate-friendly technologies and renewable sources of energy. Investments in renewable energy projects have been encouraged as a result of the creation of renewable energy targets, feed-in tariffs, tax incentives, and carbon pricing mechanisms, which have also helped the transition towards a low-carbon economy. A culture of environmental responsibility and sustainable development has been fostered as a result of policy initiatives that promote energy efficiency standards, green building codes, and sustainable procurement practices. These initiatives have also contributed to the mainstreaming of climate-friendly technologies and practices across various sectors.

In addition, the formation of public-private partnerships and international collaborations has facilitated the sharing of knowledge, the transfer of technology, and the building of capacity in the field of renewable energy and climate-friendly technology. This has fostered global cooperation and collective action in the face of the challenges posed by climate change and the degradation of the environment. The incorporation of sustainability considerations into the decision-making processes of financial institutions and investment firms has also encouraged responsible and ethical investment practices. This has resulted in the redirection of capital toward environmentally sustainable and socially responsible projects, which has facilitated the transition toward a global economy that is more resilient and sustainable.

The Obstacles and Potential for the Future:

In spite of the great progress that has been made in the field of renewable energy and technology that is friendly to the environment, the pursuit of a sustainable and resilient future still presents a number of obstacles as well as opportunities. Researchers, policymakers, and industry stakeholders continue to focus a significant amount of attention on one particular sector because of the intermittent nature of renewable energy sources. At the same time, there is a pressing need for trustworthy energy storage solutions and system modernization. The creation of energy storage technologies that are capable of fulfilling the growing demand for energy that is both clean and reliable constitutes an essential sector in which innovation and investment are needed. These technologies must be cost-effective and scalable.

In addition, the electrification of a variety of industries, such as heating and cooling, as well as industrial processes, gives a potential to cut emissions of greenhouse gases and improve energy efficiency. In order to decarbonize industries that have traditionally relied on fossil fuels, innovative technology such as heat pumps, electric boilers, and energy-efficient industrial processes can play a vital role. This further aligns with the shift to renewable energy and sustainable practices.

Another topic that should receive emphasis is the environmentally responsible administration of resources such as land, water, and raw materials. The creation of models for a circular economy, which place a priority on resource efficiency, waste reduction, and material recycling, can contribute to mitigating various industries' negative effects on the environment and reducing their carbon footprints. Innovations in materials science, such as biodegradable plastics and sustainable construction materials, play a significant role in the promotion of a resource-management strategy that is more sustainable and less harmful to the environment.

In addition, the development of environmentally friendly technologies and renewable sources of energy is intricately linked to socioeconomic and behavioral shifts that are taking place. Education, increased knowledge, and the active participation of individuals and communities are vital components in the process of developing a culture of environmental stewardship and promoting the adoption of sustainable practices. Important parts of the shift toward a society that is more ecologically conscientious and sustainable include the adoption of climate-friendly activities into everyday life, sustainable consumerism and responsible consumer choices, and the integration of renewable energy sources.

The worldwide commitment to combating climate change, as shown by international accords such as the Paris Agreement, acts as a potent catalyst for driving further developments in renewable energy and technology that is friendly to the environment. This commitment is exemplified by international agreements such as the Paris Agreement.

The demand for creative solutions is expected to continue increasing as countries collaborate to meet their emission reduction goals and transition to economies with lower levels of carbon emissions. This collaborative effort not only promotes technological improvements, but it also helps to strengthen the worldwide consensus on the necessity of promoting environmental sustainability and reducing the effects of climate change.

In order to effectively address the urgent issues of climate change, environmental degradation, and energy sustainability, technological advancements in the fields of renewable energy and climate-friendly technologies are essential. Renewable energy sources have seen a substantial expansion in both their scope and potential as a result of the development of technology such as solar panels with high efficiency, advanced technologies for wind turbines, and advancements in bioenergy and energy storage systems. In addition, the implementation of technology for smart grids, sustainable transportation solutions, and policy frameworks that are supportive has produced an atmosphere that is conducive to the adoption of climate-friendly technologies and activities.

The continued innovation and commitment to sustainability offer a promising road towards a more resilient and environmentally responsible future. This is despite the fact that problems still exist, such as the intermittent nature of renewable energy sources and the requirement for sustainable resource management. The transition to a more sustainable and low-carbon global economy will require the combined efforts of governments, industries, and communities all over the world, as well as the active participation of individuals in the adoption of climate-friendly behaviors. This will be essential in order to successfully drive the shift. Not only are innovations in renewable energy and climate-friendly technologies driving the future of energy, but they are also shaping the future of our planet. These innovations are paving the way for a world that is more sustainable and wealthy.

11.3 The role of individual and collective action in combating global warming

There is a huge and immediate danger that our planet faces as a result of global warming, which is caused by the increasing concentration of greenhouse gases in the atmosphere of the Earth. It has far-reaching repercussions, including the rise in global temperatures and the occurrence of extreme weather events, as well as the rise in sea level and disturbances in ecosystems. In order to effectively respond as a community to this catastrophe, we need to take a multi-pronged approach that incorporates both individual and group efforts. This article examines the essential roles that individuals and communities play in the battle against global warming. It looks at the ways in which individuals and communities may help to reducing the effects of global warming and supporting sustainable practices.

Comprehension of One's Own Personal Carbon Footprint

One of the most important things that individuals can do to fight global warming is to cut down on the amount of carbon that they emit into the atmosphere. A person's lifestyle and activities, such as their energy consumption, mode of transportation, nutrition, and consumption patterns, all contribute to the overall amount of greenhouse gas emissions that are linked with that person's "carbon footprint." It is vital to gain an understanding of one's own carbon footprint in order to effectively lessen this footprint through the decisions one makes.

Effective Use of Energy

Increasing energy efficiency in one's lifestyle and house is one of the most effective strategies for individuals to reduce the amount of carbon dioxide emissions they are responsible for. This encompasses a wide range of practices, including the installation of energy-efficient equipment, the insulation of homes, the prevention of drafts, and the reduction of energy usage that is not essential. Reducing one's dependency on fossil fuels and minimizing one's overall carbon footprint can be accomplished in a number of productive ways, one of which is by making the switch to renewable energy sources such as solar or wind power.

Transportation That Is Ecologically Sound

Because transportation is responsible for a sizeable portion of the world's carbon emissions, adopting environmentally responsible modes of transportation is essential to the fight against global warming. A large reduction in the carbon emissions caused by individual travel can be achieved by choosing modes of transportation such as public transportation, carpooling, biking, or walking. In addition, the proliferation of public transportation infrastructure and the use of electric vehicles both contribute to the development of a transportation system that is more sustainable and less harmful to the environment.

Consumption with Responsibility

The adoption of more responsible consuming habits is another way in which individuals can contribute to good change. This includes being conscious of the environmental impact of the goods and services purchased, selecting solutions that are sustainable and kind to the environment, cutting down on the amount of trash produced, and lending financial support to companies whose primary focus is the preservation of the natural environment. It is possible to make a considerable contribution to lowering one's overall carbon footprint by adopting sustainable activities such as recycling, reusing, and reducing the amount of single-use plastics.

The Importance of Participation in Group Efforts

While the efforts of individuals are significant, collective action is even more important in the fight against global warming. Communities, organizations, and governments all play an important part in the implementation of policies and initiatives that promote environmental protection and sustainability.

Participation in the Community and Lobbying

Local communities have the ability to be agents of change by banding together in coordinated efforts to increase awareness, promote sustainable behaviors, and advocate for environmental legislation at the local level. A culture of environmental responsibility and public participation in climate action can be cultivated through community-based activities such as tree-planting drives, local clean-up efforts, and educational programs on sustainable living.

The Responsibility of Businesses

Businesses and corporations have a substantial influence on the total amount of carbon emissions produced worldwide. It is possible to contribute to a reduction in the overall carbon footprint of commercial activities by embracing corporate social responsibility and implementing sustainable business practices. Some examples of sustainable business practices include lowering energy usage, decreasing trash output, and investing in renewable energy sources. In addition, bolstering and investing in environmentally friendly technology and sustainable innovations has the potential to generate good change across a variety of businesses and sectors.

Policies of the Government and International Cooperative Efforts

When it comes to tackling the issue of climate change on a global scale, the role of government policy and international cooperation is absolutely essential. It is possible to foster sustainable growth and make the transition to a low-carbon economy through the implementation of policy measures such as carbon pricing mechanisms, subsidies for renewable energy, and environmental restrictions. In addition, international collaborations and agreements, such as the Agreement of Paris, serve as key foundations for global cooperation in the fight against global warming and the promotion of environmentally sustainable practices.

Investing in the Infrastructure of Renewable Energy Sources

Taking collective action also requires making investments in the creation and growth of infrastructure for renewable energy sources. Renewable energy technologies, such as solar, wind, hydroelectric, and geothermal power, can be supported through research, development, and deployment efforts by governments, commercial businesses, and international organizations working together. The acceleration of the transition away from the use of fossil fuels and the contribution to a more sustainable energy future can be contributed to by the promotion of clean energy programs and the integration of renewable energy into existing power grids.

Education and Scientific Investigation

When it comes to collective action against global warming, education and research are two of the most important components.

It is possible to improve the general public's understanding of climate change and encourage the adoption of sustainable habits by making investments in environmentally focused education initiatives, scientific research, and technology innovation. It is possible to stimulate the creation of novel solutions and strategies for solving the complex challenges posed by global warming by encouraging interdisciplinary collaboration and knowledge exchange among scientists, politicians, and communities.

The fight against global warming calls for an approach that is both comprehensive and collaborative, and that takes into account activities taken both individually and collectively. Individuals have the potential to make large contributions toward the reduction of global carbon emissions if they gain an awareness of their personal carbon footprint and implement environmentally responsible behaviors in their consumption, energy use, and modes of mobility. In addition, the combined efforts of communities, businesses, and governments are necessary for the successful implementation of regulations, the promotion of innovation, and the investment in the infrastructure of renewable energy sources in order to create a future that is more sustainable and resilient. We have the ability to mitigate the effects of global warming and preserve the health and well-being of our planet for future generations if we work together and make a common commitment to environmental stewardship. This may be accomplished by collective action and a shared commitment.

Chapter 12

Looking to the Future

In light of the fact that mankind is on the verge of entering a new era that will be marked by rapid technology breakthroughs, unparalleled global difficulties, and a growing awareness of the need of sustainability and inclusion, it is necessary that an investigation into the prospects and possibilities that the future contains be conducted. This all-encompassing investigation dives into the many aspects of the future, assessing the prospective trajectories of technological, societal, and environmental advancements while also addressing the difficulties and possibilities that are still to come in the future. We are able to pro-actively prepare for the path towards a society that is more equal, sustainable, and interconnected if we imagine a future in which innovation, compassion, and foresight serve as guiding principles.

The Influence of Technological Changes on the Nature of the Future Landscape

The rate of technological innovation is accelerating at an ever-increasing rate, which is profoundly changing many elements of human existence as well as human connection. Artificial intelligence (AI), robotics, biotechnology, and quantum computing are just few of the important technical developments that are projected to play a significant role in defining the future landscape and driving extraordinary success across a wide range of industries.

The rise of automation and artificial intelligence

It is projected that artificial intelligence will change a wide variety of industries, including healthcare, banking, manufacturing, and transportation, amongst others. This revolution will be powered by machine learning algorithms and deep neural networks. It is anticipated that the incorporation of AI-driven automation would result in the simplification of procedures, the improvement of overall efficiency, and the facilitation of the development of personalized and adaptive services. Furthermore, the ethical and societal consequences of AI, which include worries about data privacy, algorithmic prejudice, and job displacement, demand proactive regulatory frameworks and ethical standards to ensure the responsible and equitable deployment of AI technologies. These frameworks and guidelines are necessary to ensure that AI technologies are deployed in a manner that is responsible and fair.

Innovations in Robotics and High-Tech Manufacturing

The rise of robotics and advanced manufacturing processes is destined to reshape the industrial landscape, leading to the development of production systems that are both highly efficient and flexible. This change is expected to occur in the near future. It is anticipated that collaborative robots, also known as cobots, would operate alongside people in a variety of manufacturing situations, thereby enhancing both productivity and safety. In addition, the introduction of additive manufacturing, which is also known as 3D printing, is poised to alter traditional production processes by enabling quick prototyping, customisation, and decentralized manufacturing capabilities. This development is set to revolutionize traditional production methods.

Both Biotechnology and Genetic Engineering are Used Here

The discipline of biotechnology is on the cusp of substantial advances in a variety of areas, including the invention of innovative biopharmaceuticals and tailored medical treatments, the development of cutting-edge tools for editing genes, and more. Genetic engineering, in conjunction with developments in genomic research and precision medicine, has the potential to change the treatment of disease as well as the prevention of disease, leading to healthcare treatments that are more individualized and successful. However, the ethical and regulatory problems associated with gene editing and genetic modification entail the necessity of robust governance frameworks as well as transparent public discourse in order to assure the responsible and ethical implementation of advances in biotechnology.

Computing on Quantum Computers and the Protection of Information

The development of quantum computing holds the potential to bring in a new era of processing capacity and the ability to solve problems. This will make it possible to rapidly simulate complicated systems, optimize algorithms, and make advances in cryptography and data security. Quantum computing has the potential to revolutionize sectors such as materials science, drug development, and climate modeling, which could lead to advances in science and technology that have never been seen before. However, in order to exploit the full potential of quantum computing while mitigating the hazards connected with it, ongoing research efforts and collaborative activities are required. These efforts are necessitated by the difficulties associated with constructing scalable and error-correcting quantum systems, as well as the consequences for data privacy and security.

Managing Changing Demographics and the Dynamics of Culture in the Context of Societal Evolution

It is anticipated that the future social environment will go through significant shifts, which will be driven by shifting demographics, evolving cultural dynamics, and the requirement of ensuring social inclusion and equity. It is projected that the social fabric of the future will be shaped in significant ways by a number of fundamental developments, including population transitions, urbanization, cultural diversity, and the redesigning of educational and employment patterns.

Alterations in the Population's Demographics and Their Aging

It is projected that global demographic changes, such as an aging population and shifts in population distribution, will cause societal structures to be rethought and will compel the development of innovative solutions in order to meet the changing requirements of a variety of age groups. It is anticipated that the growing proportion of senior people will increase demand for age-friendly infrastructure, healthcare services, and social support systems. This highlights the significance of encouraging initiatives that stimulate active and healthy aging as well as intergenerational solidarity.

Urbanization and Creating Cities That Are Sustainable

It is anticipated that the existing trend of urbanization will continue, with a greater proportion of the world's population dwelling in urban centers. This trend highlights the necessity of constructing sustainable and resilient cities that promote fair access to critical services, public transit that is both efficient and effective, green spaces, and affordable housing. It is vital to incorporate sustainable urban planning methods and smart city technology into the planning of urban areas in order to reduce negative impacts on the environment, improve quality of life, and encourage inclusive and accessible urban settings for all citizens.

Diversity in Culture and Affirmation of Belonging in Society

A recommitment to developing social inclusion, embracing cultural pluralism, and promoting intercultural discussion and understanding is required because of the growing cultural diversity and interconnectivity of countries. In order to cultivate a global community that is more accepting of differences and more conducive to harmony, it is essential to make efforts to honor cultural history, to encourage multilingual education, and to provide venues for cross-cultural collaboration and exchange. Building a society that is more compassionate and equitable is impossible without first addressing the inequities that are embedded in the system, working toward social justice, and protecting the rights of communities that are disadvantaged and at a disadvantage.

Rethinking the Way Work and Education Are Done

It is anticipated that the future of education and employment would be defined by increased flexibility, increased digital integration, and a trend toward learning and skill development that occurs throughout one's entire life. The emergence of online learning platforms, immersive educational technologies, and personalized learning experiences is set to redefine traditional education paradigms. This will make it possible for individuals from all walks of life and socioeconomic backgrounds to gain access to high-quality educational and training opportunities, regardless of where they live or their economic situation. In a similar vein, the proliferation of remote work, flexible employment models, and the gig economy highlights the necessity for adaptable labor policies, social protection mechanisms, and continuous upskilling initiatives. These are necessary in order to ensure the resilience and well-being of the workforce in the face of changing job market dynamics and technological disruptions.

Environmental sustainability: laying the groundwork for a more environmentally friendly and robust future

The protection of the ecosystems of the globe and the slowing down of climate change will continue to be essential imperatives for the purpose of ensuring the health and well-being of both the current generation and the generations to come. Climate action, conservation efforts, adoption of renewable energy models, and circular economy models are some of the key initiatives and strategies that are poised to drive the global sustainability agenda as the world grapples with the urgent need to transition to sustainable practices and foster environmental stewardship. Several key initiatives and strategies are poised to drive the global sustainability agenda.

Building Resilience While Addressing Climate Change

In order to effectively battle the growing dangers posed by climate change, it is imperative that climate action be sped up and that comprehensive measures for adaptation and mitigation be put into place. This comprises the reduction of greenhouse gas emissions, the promotion of renewable energy sources, the strengthening of climate resilience in vulnerable populations, and the deployment of nature-based solutions to lessen the impact of extreme weather events and natural disasters. All of these are necessary steps. In addition, the incorporation of climate considerations into policy frameworks, the development of infrastructure, and business strategies is essential to the process of developing a global economy that is more sustainable and robust to the effects of climate change.

Restoration of ecosystems and protection of biological diversity

The protection of the planet's natural heritage and the preservation of its ecosystems are dependent on the conservation of biological diversity and the restoration of ecosystems, respectively. The preservation of ecosystems in their entirety and the promotion of biodiversity conservation are both dependent on the implementation of programs that have as their primary objectives the preserving of endangered species, the promotion of sustainable land management practices, and the combating of deforestation and the loss of habitat. Fostering sustainable environmental management practices and preserving the rich biodiversity of the world for future generations requires significant teamwork on the part of local communities, non-governmental organizations, and governments.

Transition to renewable energy sources and environmentally responsible infrastructure

The global sustainability agenda includes as essential pillars the shift to renewable energy sources and the promotion of sustainable infrastructure development. Adopting renewable energy sources, such as solar, wind, hydroelectric, and geothermal power, is essential to lowering dependency on fossil fuels, reducing carbon emissions, and fostering the growth of an energy sector that is more sustainable and robust.

In addition, the incorporation of green building methods, the growth of public transit networks, and the integration of environmentally friendly technologies into urban planning and infrastructure projects are all essential to the development of a more ecologically responsible and sustainable future.

Models of a Circular Economy and Considerations for Sustainable Consumption

The implementation of models of a circular economy is gaining popularity as a means of maximizing the efficient use of resources, lowering the amount of waste produced, and minimizing the negative effects of production and consumption on the environment. In order to achieve sustainable and responsible resource management, it is essential to adhere to the principles of the circular economy. These principles place an emphasis on the product's durability, repairability, recycling, and minimization of waste. The transition to a circular economy can be helped along by promoting environmentally friendly product design, reducing the amount of single-use plastics used, and encouraging sustainable consumption behaviors. This will help move the needle in the direction of achieving the goals of environmental sustainability and resource conservation.

Prioritizing One's Holistic Health and Mental Wellness in the Interest of Health and Well-Being

To create a brighter future, it is essential to prioritize the overall health and happiness of individual people. It is of the utmost importance to recognize the significance of both mental and physical health, as well as the critical function of healthcare systems, preventative medicine, and mental wellness initiatives, in order to guarantee the physical and mental well-being of communities all over the world.

Integrative and Holistic Approaches to Preventive Medicine

The burden of disease can be significantly reduced and overall well-being can be significantly improved via the promotion of preventative medicine and holistic health practices. When it comes to the prevention and management of chronic health disorders, it is absolutely necessary to place an emphasis on health education, changes in lifestyle, early detection, and access to quality healthcare services. In addition, treating the mental and emotional well-being of individuals necessitates the implementation of techniques for stress management and the provision of support services for mental health within healthcare systems.

Maintaining a Healthy Mind and Body While Managing Stress

It is vital, in order to meet the issues that will be faced in the future in terms of mental health, to acknowledge mental wellbeing as an essential component of overall health. In order to address individuals' emotional well-being, it is essential to implement programs that reduce the social stigma associated with mental health concerns, broaden people's access to mental health services, and encourage the practice of stress management and the development of resiliency.

In addition, the development of communities that are caring, supportive, and that place a priority on mental health is an essential component in the process of establishing a society that is more emotionally robust and inclusive.

Preparedness for Pandemics Around the World and Global Health Security

Recent crises affecting global health have brought attention to the significance of ensuring pandemic preparation and global health security. It is necessary to establish strong public health infrastructure, early warning systems, and international coordination in disease surveillance and response if we are going to reduce the impact of infectious illnesses and protect the health of people all over the world. In addition, investments in research, the development of vaccines, and the capacity building of healthcare systems are essential for providing a coordinated and successful response to threats to global health.

Fostering a Society That Is Compassionate and Just Requires Both Inclusivity and Equity

A society that is compassionate and just must be founded upon the fundamental ideals of inclusivity and equity. Building a future that is characterized by fairness and compassion requires first and foremost an acknowledgment of the significance of issues such as social justice, gender equality, human rights, and the empowerment of populations that are disadvantaged and vulnerable.

Justice for all and respect for human rights

In order to effectively combat systematic inequities, discrimination, and injustice, the advancement of social justice and human rights is absolutely necessary. In order to cultivate a society that is more compassionate and equitable, it is essential to make efforts to eliminate discriminatory structures, work toward eliminating racial inequities, and protect human rights. Building a just and inclusive world requires a multifaceted approach that prioritizes expanding access to justice, providing assistance to underserved communities, and fighting for the civil liberties of vulnerable population groups.

Emancipation and advancement of women and girls

When it comes to establishing social inclusion and equity, gender equality and the empowerment of women and girls are two of the most important factors. For the purpose of ensuring that people of all genders have equal access to opportunities and rights, it is vital for initiatives to be undertaken with the goals of improving gender parity, promoting women in leadership positions, and combating gender stereotypes. In addition, the development of safe and welcoming environments for women, LGBTQ+ people, and members of other underrepresented groups is essential to the growth of a society that values variety and recognizes the importance of inclusiveness.

Engagement and Participation within the Community

In order to construct a society that is compassionate and just, it is essential to have individuals and communities actively engage in and participate in the building process. It is essential to encourage civic participation, grassroots activism, and community-led projects if one wishes to magnify the voices of individuals and develop collective action for the purpose of bringing about social change. In addition, making chances for communication, collaboration, and the co-creation of solutions with communities is essential to addressing the specific requirements and concerns that are important to them.

The Role of International Cooperation in the Formation of Partnerships for the Future of the World

The opportunities and problems of the future are intrinsically global in nature, and as a result, they require cooperation, diplomacy, and collaboration on an international scale. It is necessary, in order to address common global concerns and develop a society that is more integrated and peaceful, to recognize the significance of diplomacy, peacebuilding, humanitarian relief, and global governance.

Diplomacy and methods of resolving conflicts

When it comes to resolving international disagreements, encouraging peaceful cohabitation, and preventing hostilities, the employment of diplomacy and other techniques for conflict resolution are essential components. It is crucial to maintaining global peace and avoiding the human and economic toll of protracted conflicts to encourage dialogue, negotiations, and diplomatic solutions to global crises, including territorial disputes, political conflicts, and trade difficulties.

Constructing Peace and Achieving Reconciliation

In order to maintain long-term stability and societal healing, it is vital to engage in peacebuilding initiatives. These activities seek to promote reconciliation while also rebuilding societies that have been torn apart by violence. In order to establish resilient and peaceful societies, it is essential to undertake initiatives that address the underlying causes of wars, provide support for the reconstruction of areas affected by war, and involve local populations in attempts to achieve reconciliation and reintegration.

Aid to those in need and solidarity with people all throughout the world

When reacting to humanitarian crises, such as natural disasters, conflict-induced displacement, and public health emergencies, humanitarian aid and international solidarity are essential components of an effective response. The provision of humanitarian aid, the protection of vulnerable populations, and the promotion of global solidarity and burden-sharing are essential components in the effort to alleviate human suffering and ensure that the fundamental requirements of communities that have been impacted are addressed.

Governance on a Global Scale and Multilateralism

In order to effectively solve the global difficulties that we all face, it is essential to engage in cross-national collaboration and democratic decision-making, as emphasized by the concepts of multilateralism and global governance. When it comes to coordinating solutions to global crises such as climate change, public health emergencies, and economic stability, having effective multilateral institutions, international agreements, and collaborative initiatives are vital components. In order to construct a global society that is more integrated and robust, it is essential to acknowledge the significance of multilateralism and to encourage the development of diplomatic solutions to global crises.

When we consider the future, we are met with a patchwork of difficulties, opportunities, and revolutionary possibilities. These possibilities include technological innovation, societal evolution, environmental sustainability, health and well-being, inclusivity and equity, and international cooperation. In addition, the future presents us with a tapestry of challenges, opportunities, and transformative possibilities. To be able to imagine a future that is shaped by innovation, compassion, and foresight, we need to make a concerted effort to face the global issues that lie ahead while also capitalizing on the possibilities for good change. We can proactively prepare for the journey towards a more equitable, sustainable, and interconnected world by forging a vision of the future that is marked by inclusivity, sustainability, and social justice. In this world, the potential for progress is limitless, and the capacity for compassion knows no bounds.

12.1 Anticipating the long-term consequences of melting ice and rising seas

One of the most important and difficult concerns of our time is represented by the effects that melting glaciers and increasing sea levels will have on the environment. The polar ice caps and glaciers are melting at an accelerated rate, which is accelerated by global warming and climate change. This melting has far-reaching ramifications for both natural ecosystems and human cultures. Understanding the scope of the problem and devising effective measures to minimize and adapt to the effects of melting ice and rising seas are two of the most important reasons why it is crucial to make predictions about the long-term repercussions of the changes that are occurring.

Influence on the Climate and Weather Patterns of the Whole World

The melting of ice caps and glaciers adds to changes in global climate patterns, which in turn has an effect on weather systems and creates a feedback loop that makes the planet warm up even more. As ice melts, the reflecting surface of the polar areas reduces, which leads to an increase in the amount of solar radiation absorbed by the surface of the Earth, which in turn leads to additional warming of the surface of the planet.

This mechanism, which is known as the ice-albedo feedback, not only quickens the rate at which ice is melting, but it also messes up the patterns of air circulation. As a result, extreme weather events such as storms, hurricanes, and heatwaves occur more frequently and with greater intensity.

Sea Level Rise and the Erosion of Coastal Lands

The rise in sea levels around the world is one of the most severe impacts that can result from glacier melting. The melting of glaciers and polar ice caps leads to an influx of water into the oceans, which causes sea levels to gradually rise. This poses a direct threat to coastal populations and low-lying places all over the world. This phenomena makes coastal erosion, floods, and the entry of seawater worse, which in turn results in the loss of valuable land, damage to infrastructure, and the relocation of communities. Small island nations and densely populated coastal cities are particularly susceptible to the effects of sea-level rise, which has the potential to cause enormous economic, social, and environmental disruptions. This is especially the case in the case of small island nations.

Destabilization of Ecosystems and Loss of Biological Diversity

The melting of ice caps and glaciers has significant repercussions for the ecosystems of the arctic and alpine regions, leading to the disruption of biodiversity and the ecological balance of these regions. Polar regions are home to a wide variety of unique species that have evolved to survive in the harsh circumstances of these ecosystems. Some examples of these species include polar bears, seals, penguins, and numerous cold-adapted marine organisms. The loss of ice habitats, decreased availability to food sources, and changes in oceanic and atmospheric conditions all pose a threat to the continued existence of these species, which could result in significant population decreases and the destabilization of the ecosystem. Additionally, the disruption of polar ecosystems has cascading consequences on global biodiversity and marine food chains, which has an effect on fisheries as well as marine life and coastal economy.

Management of Water Resources and Provision of Drinking Water

Additionally, the melting of glaciers and ice caps has huge repercussions for the freshwater supplies and water supply systems across the globe. Glaciers are the source of water for many of the world's rivers, which makes glaciers an important source of freshwater for a variety of uses, including agriculture, drinking water, and industrial processes. The slow depletion of glacier-fed rivers and the interruption of water supply present issues for the management and sustainability of water resources, particularly in places that are highly dependent on glacial meltwater. These challenges are particularly acute in the western United States. In addition, the disruption of water cycles and the rise in the probability of water scarcity make the problems that already exist with regard to water management, agricultural production, and the generation of hydroelectric power more worse.

Consequences for International Migration and the Displacement of People

The long-term effects of melting glaciers and rising seas have enormous implications not just for the migration patterns of humans but also for the relocation of communities. Many communities are being forced to face the realities of relocating and adapting to new living settings as a consequence of the threat posed by rising sea levels and flooding along the coast. This phenomena not only imposes major social, economic, and psychological pressures on people that are being affected by it, but it also creates difficulties for the preservation of cultural legacy and the identity of communities. The likelihood of migration and displacement as a result of climate change highlights the critical need of implementing comprehensive plans for the adaptation to climate change, reduction of the risk of natural disasters, and protection of human rights for people that are particularly vulnerable.

Impacts on the Socioeconomic System and the Vulnerability of the Infrastructure

The socio-economic effects of melting ice and rising seas go beyond environmental concerns to include serious problems for the durability of infrastructure and for the economy's overall stability. Significant economic losses and interruptions in global trade and transportation can be attributed to the increased vulnerability of coastal infrastructure, which includes ports, roadways, and urban centers, to the effects of sea-level rise, storm surges, and extreme weather events. In addition, the consequences for property prices, insurance costs, and the long-term survival of coastal economies highlight the necessity for proactive planning, investments in resilient infrastructure, and the incorporation of climate adaption strategies into urban development and land-use planning.

Cooperation on a global scale and the development of climate-resilient strategies

Because of the complexity of the effects that will result from melting glaciers and rising sea levels, it will be necessary to make efforts that are all-encompassing and include collaboration on a global, regional, and local scale. When it comes to addressing the underlying factors that contribute to global warming, lowering emissions of greenhouse gases, and promoting sustainable development practices, international collaboration and coordinated action are absolutely necessary. In addition, the development of climate resilience plans, disaster preparedness programs, and adaptive measures for communities that are susceptible is essential for protecting the well-being and livelihoods of impacted populations, reducing the negative effects of sea-level rise, and limiting the negative effects of sea-level rise.

The funding of climate research and monitoring systems

It is absolutely necessary to make investments in climate science, monitoring systems, and research programs in order to improve our understanding of the complex dynamics of ice melting and sea levels rising. Insights into the dynamics of polar ice melt, sea-level rise, and the connections between the cryosphere and the global climate system can be gained through the establishment of comprehensive monitoring networks, satellite-based observations, and modeling techniques.

In addition, the promotion of multidisciplinary collaborations and the exchange of information among scientists, policymakers, and other interested parties is essential to the development of effective strategies for mitigating the long-term effects of melting ice and rising sea levels and adapting to these changes.

Understanding the magnitude of the issues faced by global warming and climate change requires having a thorough understanding of the long-term effects that can be anticipated from the melting of ice and the rising of sea levels. The consequences of melting ice and rising seas require proactive measures and collaborative efforts to promote climate resilience, environmental sustainability, and social equity. From the impacts on global climate patterns and coastal ecosystems to the socio-economic implications for vulnerable communities and infrastructure, the consequences of melting ice and rising seas necessitate such measures and efforts. We can work towards a more resilient and sustainable future by giving higher priority to investments in climate science, fostering international cooperation, and implementing comprehensive climate adaptation strategies. This will allow us to ensure that the preservation of our planet's cryosphere and the protection of vulnerable communities are given the utmost importance.

12.2 The urgency of addressing global warming for future generations

Future generations face a grave and immediate risk to their health and their capacity to maintain their standard of living as a result of global warming, which is being caused by the rising concentration of greenhouse gases in the atmosphere of the Earth. It is critical that the urgent need for concerted global action to mitigate the effects of climate change and protect the health and vitality of our planet for future generations be recognized as soon as possible. The repercussions of climate change are becoming more and more apparent. This all-encompassing investigation dives into the myriad facets of the pressing need to address global warming. It investigates the environmental, social, economic, and ethical imperatives for taking prompt and decisive action in order to protect the future of humanity and the world.

Concerns Regarding the Environment: Protecting Ecosystems and Biological Diversity

The severe environmental repercussions of global warming, which pose a threat to the integrity of ecosystems and the biodiversity of the globe, highlight the critical nature of tackling this issue as quickly as possible. The effects of increasing global temperatures, varying weather patterns, and the destruction of natural habitats have far-reaching repercussions for plant and animal species. These factors contribute to the loss of habitat, the extinction of species, and the deterioration of ecosystems. Because the extinction of critical species and ecosystems can disrupt food chains, diminish ecosystem resilience, and undermine the planet's ability to support life, the ramifications for global biodiversity and ecological balance are enormous. It is imperative that urgent action be taken to address global warming if we are to protect the incredible variety of life on Earth and ensure the long-term viability of natural ecosystems for the benefit of future generations.

Humanitarian Emergencies and Natural Catastrophes Caused by Climate Change

The rising frequency and severity of climate-related catastrophes, such as hurricanes, droughts, floods, and wildfires, underline the necessity of addressing global warming in order to decrease the risk of humanitarian crises and safeguard communities who are vulnerable to its effects. Extreme weather can have a major effect on human lives, livelihoods, and infrastructure, which can result in displaced populations, food shortages, water scarcity, and public health issues. In order to improve disaster preparedness, promote climate resilience, and provide humanitarian aid and support to communities who have been impacted by climate-related disasters, it is imperative that immediate action be taken to combat global warming. In order to ensure the health and safety of future generations in the face of the difficulties posed by climate change, preventative actions, such as the installation of early warning systems, climate-resilient infrastructure, and community-based adaptation efforts, are absolutely necessary.

Scarcity of water and uncertain food supplies

The impact of global warming on the worsening of water scarcity and food insecurity poses enormous difficulties for global food production, agricultural sustainability, and access to clean water resources. Global warming adds to the exacerbation of water shortage and food insecurity. The effect that shifting weather patterns have on crop yields, the amount of water that is available, and agricultural productivity has repercussions for food prices, nutritional security, and the livelihoods of farming people all over the world. In order to guarantee that future generations will have access to food that is both safe and nutritious, it is imperative that urgent action be taken to combat global warming. This can be accomplished by developing climate-resilient farming methods, improving agricultural resilience, and promoting sustainable water resource management. In order to effectively address the interrelated problems of water scarcity and food insecurity in the context of global warming, it is vital to make investments in environmentally responsible agriculture, methods of water conservation, and innovative agricultural practices.

Concerns for Personal Health as Well as the General Population's

Warming caused by human activity poses serious threats to public health, including an increased risk of heat-related illnesses, the transmission of infectious diseases, and a worsening of the symptoms of pre-existing medical disorders. The urgency with which action needs to be taken to combat global warming in order to protect the health and well-being of future generations is highlighted by the impact that rising temperatures will have on the quality of the air we breathe, the diseases that are carried by water, and the illnesses that are transmitted by vectors. When it comes to minimizing the dangers posed by global warming and protecting the health of vulnerable people, the development of community-based health programs, the implementation of public health interventions, and the promotion of healthcare systems that are climate-resilient are all crucial components. In order to mitigate the negative effects of climate change on public health and safeguard the wellbeing of future generations, it is essential to place a higher priority on investments in climate-resilient infrastructure, disease surveillance, and the expansion of healthcare capacity.

Implications for the Economy: Promoting Sustainable Development and Resilience

The urgency of tackling global warming is inextricably tied to its economic implications, which have deep consequences for global development, economic stability, and the well-being of future generations. The urgency of addressing global warming is directly proportional to the magnitude of these consequences. The effects that climate change will have on several economic sectors, including as agriculture, tourism, and the construction of new infrastructure, highlight the importance of taking preventative actions to foster sustainable development and resilience in the face of the effects of global warming. In order to develop sustainable economic growth, promote green technology and solutions for clean energy, and provide chances for job creation and innovation in climate-resilient industries, addressing global warming as quickly as possible is necessary. Building a more sustainable and resilient global economy that places a priority on the well-being of future generations requires the incorporation of climate considerations into economic policies, investment plans, and corporate practices. This is one of the most important steps in this process.

Transition to renewable energies and emerging technologies

The urgency of addressing global warming is highlighted by the requirement to transition towards renewable energy sources and promote sustainable energy technology in order to minimize dependency on fossil fuels and cut carbon emissions. This transition and promotion of sustainable energy technologies is critical for two reasons. In order to enhance energy security, reduce greenhouse gas emissions, and promote environmental sustainability for future generations, the development of clean energy solutions, such as solar, wind, hydroelectric, and geothermal power, is vital. Other examples of clean energy solutions include bioenergy and nuclear power. In order to prepare the way for a more sustainable and carbon-neutral energy future, the urgency with which we must address global warming is vital in accelerating the deployment of clean energy technology, increasing energy efficiency, and supporting the adoption of clean energy laws and incentives. All of these are important steps.

The promotion of intergenerational justice and responsible environmental stewardship are ethical imperatives

The urgency of tackling global warming is founded in ethical imperatives that highlight the moral obligation to maintain the earth for future generations and promote intergenerational fairness and environmental stewardship. The urgency of addressing global warming is anchored in ethical imperatives that accentuate the moral responsibility to protect the planet for future generations. The significance of taking prompt action to minimize the effects of global warming and ensure the well-being and sustainability of the world for future generations is brought to light by the environmental ethics, sustainability, and intergenerational equity principles. These principles underline how important it is to act now. It is imperative that urgent action be taken to address global warming if we are to cultivate a culture of environmental responsibility, promote ethical decision-making, and advocate for policies that put the long-term interests of future generations ahead of the short-term gains of the current generation.

We can foster a more sustainable and compassionate approach to solving the difficulties posed by global warming and ensure the well-being of future generations if we embrace the concepts of environmental stewardship and intergenerational justice. These values are central to the concept of intergenerational justice.

In light of the significant and intricately intertwined difficulties that climate change poses for the environment, society, the economy, and ethics, it is imperative that immediate action be taken to solve the issue of global warming for the sake of future generations. We can work toward a more sustainable and resilient future that maintains the well-being and vitality of the planet for generations to come if we prioritize investments in climate resilience, environmental protection, and sustainable development. This will allow us to move towards a more sustainable and resilient future. We can pave the way for a society that is more sustainable, equitable, and thriving via collective action, international cooperation, and a shared commitment to environmental stewardship. This world will meet the demands of future generations while also preserving the legacy of environmentally sound practices.

12.3 Call to action and a hopeful vision for a sustainable future

There is an urgent need for a collective call to action in order to build an optimistic vision for a sustainable future in light of the tremendous challenges offered to humanity by environmental degradation, climate change, and global sustainability. These issues are currently being grappled with by humanity. The necessity of finding solutions to these problems stems from the realization that we are inextricably linked to the natural environment around us and from the conviction that it is our ethical obligation to preserve the physical and mental well-being of the planet for the benefit of future generations. This in-depth investigation digs into the myriad of facets that make up the call to action for sustainability. These facets include environmental conservation, climatic resilience, social equality, and ethical stewardship. We can pave the way for a world that places equal importance on the well-being of people and the earth if we accept a common vision for a sustainable future and make a commitment to revolutionary action.

The protection of the environment and the conservation of biological diversity

In order to acknowledge the inherent importance of the planet's ecosystems and the imperative to preserve and restore their natural state, it is necessary to issue a call to action for the protection of the environment and the maintenance of biological diversity. The call to action for protecting the natural legacy of our world should include the adoption of sustainable land management techniques, the promotion of reforestation and afforestation programs, and the conservation of natural ecosystems and hotspots for biodiversity. These are all essential components. The commitment to conserving the ecological balance of the earth and cultivating a more resilient and biodiverse natural environment is highlighted by activities such as the restoration of degraded ecosystems, the conservation of endangered species, and the promotion of sustainable resource management methods. These activities all take place in the natural world.

Strategies for Resilience and Adaptation in a Changing Climate

The call to action for climate resilience and adaptation methods highlights how important it is to proactively address the implications of climate change and to promote resilience in the face of various environmental challenges. In order to mitigate the dangers that are posed by extreme weather events and rising global temperatures, it is vital to make investments in infrastructure that is climate-resilient, to promote efforts that reduce the risk of natural disasters, and to adopt solutions that are based on nature. Building a society that is more robust and adaptable, one that can resist the effects of climate change and that can promote sustainable development for future generations, requires the incorporation of climate considerations into urban planning, agricultural practices, and the management of water resources. This is an essential step in the process.

Equity in social conditions and inclusive economic growth

In the call to action for social equity and inclusive development, the necessity of tackling systemic disparities and promoting social justice and human rights for all people is recognized. This is an important step toward achieving social equality and inclusive development. It is vital to provide higher priority to investments in education, healthcare, and social protection programs in order to alleviate poverty, develop equal opportunity, and promote economic growth that is inclusive of all people. The dedication to constructing a more fair and compassionate society that values the dignity and well-being of every individual is highlighted by the empowerment of marginalized populations, the advancement of gender equality, the promotion of cultural diversity and inclusion, and the promotion of cultural diversity and inclusion.

Stewardship Based on Ethical Principles and Responsible Administration

The call to action for ethical stewardship and responsible governance places a strong emphasis on the significance of cultivating ethical decision-making, promoting transparent government, and preserving the ideals of integrity and accountability in all levels of society. In order to cultivate a culture of ethical stewardship and responsible leadership, it is vital to engage in ethical business practices, advocate for corporate social responsibility, and promote transparent policymaking. The dedication to establishing a more sustainable and just world, one in which ethical values direct every decision and action, is underscored by the commitment to preserving ethical standards, cultivating integrity, and promoting environmental and social accountability.

A Positive Perspective on the Possibility of a Sustainable Future

The shared dedication to promoting environmental harmony, social well-being, and ethical responsibility is one of the defining characteristics of a positive vision for a future that is environmentally sustainable. It envisions a world in which communities live in harmony with nature, in which the ecosystems of the Earth thrive, and in which every human has access to clean air and water as well as the resources necessary to live a life that is both dignified and successful.

This vision acknowledges the importance of social fairness, ethical stewardship, and environmentally sustainable practices as the foundational pillars around which global collaboration and transformative action are built. It cultivates a culture of collaboration, innovation, and compassion, one in which the well-being of people as well as the earth is prioritized, and one in which the heritage of sustainability is perpetuated for future generations.

A call to action for a sustainable future is a demonstration of our shared responsibility to save the earth, advance social fairness, and defend ethical standards for the sake of the well-being of both the current generation and those to come in the future. We may pave the way for a more sustainable and compassionate world that supports the harmonious coexistence of all living creatures by adopting the concepts of environmental conservation, climatic resilience, social equity, and ethical stewardship. By doing so, we can make the world a better place for all living beings. We can work towards a hopeful vision for a sustainable future that upholds the values of environmental stewardship, social justice, and ethical integrity for the betterment of humanity and the preservation of the planet through transformative action, inclusive collaboration, and a shared commitment to sustainability. This will allow us to make progress toward our goal of achieving sustainability.